Test-Driven Development in Go

A practical guide to writing idiomatic and efficient Go tests
through real-world examples

Adelina Simion

BIRMINGHAM—MUMBAI

Test-Driven Development in Go

Group Product Manager: Gebin George

Publishing Product Manager: Pooja Yadav

Content Development Editor: Rosal Colaco

Technical Editor: Jubit Pincy

Copy Editor: Safis Editing

Project Coordinator: Deeksha Thakkar

Proofreader: Safis Editing

Indexer: Sejal Dsilva

Production Designer: Prashant Ghare

Developer Relations Marketing Executive: Sonia Chauhan and Rayyan Khan

First published: March 2023

Production reference: 2190623

Published by Packt Publishing Ltd.

Livery Place

35 Livery Street

Birmingham

B3 2PB, UK.

ISBN 978-1-80324-787-8

www.packtpub.com

To my family and my partner, who are always in my corner.

– Adelina Simion

Foreword

I consult many companies on various topics, whether architecture, data pipelines, optimization, or one of many others. And every time, I find myself writing tests, sometimes a lot of tests. Needless to say, I think tests are important.

The main factor in writing tests is the cost of error. If you are writing an internal web server for lunch orders, you can fix bugs as they appear. However, if you write software that controls railroad semaphores, you will invest more time in upfront testing.

You can never test enough: SQLite has 608 lines of tests for each line of code (`https://www.sqlite.org/testing.html`), and it still has bugs. This means you need to be smart about the time and effort you invest in writing tests. You need to know what kind of tests give you the most "value for money." It might be integration tests, end-to-end tests, or even fuzzing. Probably a combination of several of these methods. You can be data-driven by marking bugs with what kind of test would have caught them – after a while, you will know where to invest your time. Remember that there is no "one size fits all" here; every project and every team will find a different set of tests more effective for them. And things will change during a project life cycle. Make sure to update your testing priorities accordingly.

No matter how much you test, bugs will get to production. NASA has a very thorough development process (`https://www.fastcompany.com/28121/they-write-right-stuff`), and they still manage to ship bugs to Mars. However, they also triage and fix (`https://futurism.com/the-byte/nasa-mars-lander-hit-itself-shovel`) bugs on Mars, which I find mind-boggling. For you, this means metrics and logging; you need to know when and why your code fails.

I could ramble about tests for a very long time; just ask my students. But you came here to hear about the book you are about to read. My advice is, read it!

Adelina does a great job of showing you the ropes with several practical examples. She will walk you through the bigger picture of testing, explaining TDD, "red, green, refactor," and more terms, and then get into the nuts and bolts of writing tests. You will learn about the difference between `t.Error` and `t.Fatal` (I use the latter), how to write table-driven tests, and many other best practices and tools. You will even learn how to Use Docker Compose for running end-to-end tests, use external libraries that automagically create mocks, and much more. And, even though this book is about testing, you will learn a lot of Go along the way.

It shows that Adelina has a lot of experience writing tests and that she is passionate about the subject. There are many tips sprinkled throughout the book: for example, how to name tests so that test files will show next to the code they are testing. It might seem like a trivial matter, but in large code bases, this tip will save you a lot of time.

I encourage you to dive in, and as you surface from time to time, head over to your code and implement what you read.

Happy hacking!

Miki Tebeka,

Founder and CEO of 353Solutions

I'm thrilled Adelina has written this book. By reading it and following along with the examples, you'll become confident with the practices of **Test-Driven Development** (**TDD**) and will hopefully adopt it as your primary method of engineering.

The first two chapters provide a solid foundation that will give you everything you need to get started. *Chapter 3, Mocking and Assertion Frameworks,* gets stuck into practices that enable you to write tests that touch multiple systems, which is explored more in the second part of the book, starting with *Chapter 5, Performing Integration Testing.* The final part of the book, *Part 3, Advanced Testing Techniques*, covers many of the more difficult areas of testing in Go, concurrency, fuzzing, and generics.

I almost exclusively follow the TDD approach and have seen it pay dividends again and again, allowing me to write very high-quality code and get it to production sooner. Strong testing skills can make the difference between a good engineer and a great one. This is why a few friends and I have created testing libraries such as **Testify** (https://github.com/stretchr/testify) and my minimalist alternative, **Is** (https://github.com/matryer/is), and **Moq** (https://github.com/matryer/moq) for generating mocks from interfaces in Go.

When I first heard about TDD many years ago, I spat my coffee out, looked directly at the camera, and pulled a very confused face indeed. How can you test something that doesn't yet exist?

Maintaining software often costs more than the initial build, especially in successful projects that have a long life, so it makes sense to invest time and energy in making maintenance easier. Tests are one of our most powerful tools for this mission. If we get tests right and they all pass, we know our code is safe to deploy. If we find a bug, we can prove it with a test (it's important to see that test fail) before fixing the issue and adding that new test to our suite so that we never see that same bug again. Tests allow us to make bold changes with confidence. Adelina dives into this in more detail in *Chapter 7, Refactoring in Go.*

This all applies regardless of when you write the tests, so why do we write the tests first?

One big misconception with TDD is that you should write all of your tests up front, before writing any "real code." In fact, the approach should be much more iterative. Write the smallest amount of test code that moves the story along, run the test and see that it fails, and then make that test pass. Rinse and repeat. *Chapter 2, Unit Testing Essentials,* covers this nicely and gives you some practice.

The whole Web 2.0 movement in the early twenty-first century taught us that the user experience matters, perhaps more than everything else. If our interfaces are clear and slick, with an elegant design, people are more likely to use our products. It's important to realize that this applies to our code as well; we have users, and there are interfaces (not just code interfaces, but our functions, structs, method names, arguments, return types – everything somebody, including ourselves, will use).

When we write tests, we become our first user. Writing tests up front pulls the interface design considerations into sharp focus. How should people use this new thing we're writing? What do they think about the problem? How would they (or us) expect to interact with these types and methods? Starting with the tests gets us thinking about this early, and can drive us towards a simpler design.

Once you've got the happy path working, you can explore the edges. *Chapter 10, Testing Edge Cases*, provides insights into property-based testing and fuzzing, which you can use to increase the robustness of your code.

I have twice mentioned that it's important to see a test fail, and there's a very good reason for this. Consider a test that can't fail:

```
is.True(true) // true should be true
```

A test that can't fail is useless and might as well be deleted. This is a trivial example, but seeing a test fail proves to you and others that it is saying something meaningful about your code. If you write a test and make it pass before seeing it fail, how can you be sure that the test is actually interacting with the code it's testing? This isn't just theoretical; I have plenty of examples throughout my career where I have made a mistake like this, and it's worse than not having a test at all because it gave me false confidence. The first chapter of this book, *Chapter 1, Getting to Grips with Test-Driven Development*, will give you a solid understanding of the advantages of TDD.

It is, of course, possible to achieve good test coverage without TDD, but it's harder. How do you know you haven't missed something important? How are those reviewing your PR supposed to know that you've got the testing right? Without checking out the code and going through deliberately breaking things, they don't. With TDD, we do know because we saw that test fail. We know it's covered.

One side-effect of TDD that I find particularly helpful is when it gives me a TODO list. I start with a user-centric perspective, and I am then delivered a series of errors and failures from the tooling. First, I get compiler errors (*Sorry, that method doesn't exist*) and, later, I get assertion errors (*Sorry, that string result wasn't what we expected*). This guides my work and helps me focus. At the end of the day, I will often leave my morning self a failing test so I can jump right back into things. I'm faster when following TDD than without it because of the mistakes I avoid, and the clarity I have earlier in the process.

If TDD feels unusual and slow to you initially, I urge you to stick with it. You'll get better. You'll get better at designing software, and as Chris James (author of *Learn Go with Tests*: `https://quii.gitbook.io/learn-go-with-tests/`) points out, TDD is the feedback loop for validating your design.

I hope you enjoy this book and your journey through TDD with Adelina. She will walk you through it all in detail, taking you deep into the rationale behind the approach and giving you lots of practical and actionable advice along the way.

I'd love to hear about your experiences. Please tweet me, `@matryer`, and share your perspective.

Mat Ryer

Engineering director at Grafana Labs

Contributors

About the author

Adelina Simion is a technology evangelist at Form3. She is a polyglot engineer and developer relations professional, with a decade of technical experience at multiple start-ups in London. She started her career as a Java backend engineer, converted later to Go, and then transitioned to a full-time developer relations role. She has published multiple online courses about Go on the LinkedIn Learning platform, helping thousands of developers upskill with Go. She has a passion for public speaking, having presented on cloud architectures at major European conferences. Adelina holds an M.Sc. in mathematical modeling and computing.

I want to thank all those who have supported me in this project, especially my technical reviewers, Stuart Murray and Dimas Prawira, without whom this project would not have been possible.

About the reviewer

Stuart Murray is an engineer with 10 years of experience across Go, Rust, Java, TypeScript, and Python. He has worked in a variety of industries including fintech, healthtech, insurtech, and marketing.

Dimas Yudha Prawira is a Go backend engineer, speaker, tech community leader, and mentor with a love for all things Go, open source, and software architecture. He spends his days developing Go microservices, new features, observability, improved testing, and best practices. Dimas explores thought leadership avenues, including reviewing Go textbooks, speaking at the Go community, and leading the Go and Erlang software engineer community. Dimas holds a master's degree in Digital Enterprise Architecture from Pradita University and a bachelor's degree in Information Technology from UIN Syarif Hidayatullah.

In his spare time, he likes to contribute to open source projects, read books, watch movies or play with his kids.

I'd like to thank my caring, loving, supportive wife, my deepest gratitude. Your encouragement when the times got rough is much appreciated.

A heartfelt thanks for the comfort and relief of knowing that you were willing to manage our household activities while I focused on my work. To my kids Khaira, Farensa, and Salah thank you and I love you.

To the memory of my father, who always believed in me. You are gone but your belief in me has made this journey possible.

Lastly, to all my friends, thank you for all the support and encouragement you give me, and the patience and unwavering faith you have in me.

Table of Contents

3

Mocking and Assertion Frameworks   65

4

Building Efficient Test Suites   91

Part 2: Integration and End-to-End Testing with TDD

5

Performing Integration Testing   123

6

End-to-End Testing the BookSwap Web Application   147

Part 3: Advanced Testing Techniques

10

Testing Edge Cases   261

11

Working with Generics   281

Preface

At the beginning of my career as a software engineer, I was focused on understanding technical concepts and delivering functionality as fast as I could. As I advanced in my career and matured my code-writing craft, I started to understand the importance of code quality and maintainability. This is especially important for Go developers since the language is designed around the values of efficiency, simplicity, and safety.

This book aims to provide you with all the tools you need to elevate the quality of your own Go code, through the industry-standard development methodology of **Test-Driven Development** (TDD). It provides a comprehensive introduction to the principles and practices of TDD, helping you get started without any prior knowledge. It also demonstrates how to apply this methodology to Go, which continues to gain popularity as a development language.

Throughout this book, we will explore how to leverage the benefits of TDD demonstrated with a variety of code examples, including building a demo REST API. This practical approach will teach you how to design testable code and write efficient Go tests, using the standard testing library as well as popular open source third-party libraries in the Go development ecosystem.

This book introduces the practices of TDD and teaches you how to use them in the development of Go applications using practical examples. It demonstrates how to leverage the benefits of TDD in applications at every level, ensuring that they deliver functional and non-functional requirements. It also touches on important principles of how to design and implement testable code, such as containerization, database integrations, and microservice architectures.

I hope you will find this book helpful in your journey to becoming a better engineer. In its pages, I have included all the knowledge that I wish I had when I first started out with Go development, which I hope will help make writing well-tested code easier for you.

Happy reading!

Who this book is for

This book is aimed at developers and software testing professionals, who want to deliver high-quality and well-tested Go projects. If you are just getting started with TDD, you will learn how to adopt its practices in your development process. If you already have some experience, the code examples will help you write more efficient testing suites and teach you new testing practices.

What this book covers

Chapter 1, Getting to Grips with Test-Driven Development, introduces the principles and benefits of TDD, setting motivation for continuing to learn about it.

Chapter 2, Unit Testing Essentials, teaches us the essential knowledge for beginning our journey with test writing. It covers the test pyramid and how to write unit tests with Go's standard testing library, and how to run the tests in our projects.

Chapter 3, Mocking and Assertion Frameworks, builds upon the knowledge from previous chapters and teaches us how to write tests for code that has dependencies. It covers the usage of interfaces, how to write better assertions, and the importance of generating and using mocks to write tests with isolated scope.

Chapter 4, Building Efficient Test Suites, explores how to group tests into test suites (which cover a variety of scenarios) using the popular Go testing technique of table-driven design.

Chapter 5, Performing Integration Testing, expands the scope of the tests we write to include the interactions between components using integration testing. It also introduces **Behavior-Driven Development (BDD)**, which is an extension of TDD.

Chapter 6, End-to-End Testing the BookSwap Web Application, focuses on building the REST API application, which is the main demonstration tool of the book. It covers containerization using Docker, database interactions, and end-to-end testing.

Chapter 7, Refactoring in Go, discusses techniques for code refactoring, which is a significant part of the development process. It covers how the process of changing dependencies is facilitated by the use of interfaces and the common process of splitting up monolithic applications into microservice architectures.

Chapter 8, Testing Microservice Architectures, explores the testing challenges of microservice architectures, which change at a rapid pace. It introduces contract testing, which can be used to verify the integration between services.

Chapter 9, Challenges of Testing Concurrent Code, introduces Go's concurrency mechanisms of goroutines and channels, including the challenges of verifying concurrent code. It also explores the usage and limitations of the Go race detector.

Chapter 10, Testing Edge Cases, expands the testing of edge cases and scenarios by making use of fuzz testing and property-based testing. It also explores code robustness, which allows us to write code that can handle a variety of inputs.

Chapter 11, Working with Generics, concludes our exploration of TDD in Go by exploring the usage and testing of generic code. It also discusses how to write table-driven tests for generic code, as well as how to create custom test utilities.

To get the most out of this book

You will need a version of Go later than 1.19 installed on your computer. All code examples have been tested using Go 1.19 on macOS. After *Chapter 6*, running the BookSwap demo application will require you to have PostgreSQL installed or run it using Docker.

Software covered in the book	Operating system requirements
Go 1.19	Windows, macOS, or Linux
PostgreSQL 15	Windows, macOS, or Linux
Docker Desktop 4.17	Windows, macOS, or Linux
Postman 10 (optional)	Windows, macOS, or Linux

The GitHub repository describes the configuration required for running the BookSwap application locally, which includes setting some local environment variables.

If you are using the digital version of this book, we advise you to type the code yourself or access the code from the book's GitHub repository (a link is available in the next section). Doing so will help you avoid any potential errors related to the copying and pasting of code.

You will get the most out of reading this book if you are already familiar with the fundamentals and syntax of Go. If you are completely new to Go, you can complete a tour of Go here: `https://go.dev/tour/list`.

Download the example code files

You can download the example code files for this book from GitHub at `https://github.com/PacktPublishing/Test-Driven-Development-in-Go`. If there's an update to the code, it will be updated in the GitHub repository.

We also have other code bundles from our rich catalog of books and videos available at `https://github.com/PacktPublishing/`. Check them out!

Download the color images

We also provide a PDF file that has color images of the screenshots and diagrams used in this book. You can download it here: `https://packt.link/KFZWx`.

Conventions used

There are a number of text conventions used throughout this book.

`Code in text`: Indicates code words in text, database table names, folder names, filenames, file extensions, pathnames, dummy URLs, user input, and Twitter handles. Here is an example: "The Go toolchain provides the single `go test` command for running all the tests that we have defined."

A block of code is set as follows:

```go
func (e *Engine) Add(x, y float64) float64{
    return x + y
}
```

When we wish to draw your attention to a particular part of a code block, the relevant lines or items are set in bold:

```go
func TestAdd(t *testing.T) {
    e := calculator.Engine{}
    x, y := 2.5,3.5
    want := 6.0
    got := e.Add(x,y)
    if got != want {
    t.Errorf("Add(%.2f,%.2f) incorrect, got: %.2f, want: %.2f",
x, y, got, want)
    }
}
```

Any command-line input or output is written as follows:

```
$ go test -run TestDivide ./chapter04/table -v
```

Bold: Indicates a new term, an important word, or words that you see onscreen. For instance, words in menus or dialog boxes appear in **bold**. Here is an example: "Select **System info** from the **Administration** panel."

> **Tips or important notes**
> Appear like this.

Get in touch

Feedback from our readers is always welcome.

General feedback: If you have questions about any aspect of this book, email us at `customercare@packtpub.com` and mention the book title in the subject of your message.

Errata: Although we have taken every care to ensure the accuracy of our content, mistakes do happen. If you have found a mistake in this book, we would be grateful if you would report this to us. Please visit `www.packtpub.com/support/errata` and fill in the form.

Piracy: If you come across any illegal copies of our works in any form on the internet, we would be grateful if you would provide us with the location address or website name. Please contact us at `copyright@packt.com` with a link to the material.

If you are interested in becoming an author: If there is a topic that you have expertise in and you are interested in either writing or contributing to a book, please visit `authors.packtpub.com`.

Share Your Thoughts

Once you've read *Test-Driven Development in Go*, we'd love to hear your thoughts! Scan the QR code below to go straight to the Amazon review page for this book and share your feedback.

https://packt.link/r/1-803-24787-8

Your review is important to us and the tech community and will help us make sure we're delivering excellent quality content.

Download a free PDF copy of this book

Thanks for purchasing this book!

Do you like to read on the go but are unable to carry your print books everywhere?

Is your eBook purchase not compatible with the device of your choice?

Don't worry, now with every Packt book you get a DRM-free PDF version of that book at no cost.

Read anywhere, any place, on any device. Search, copy, and paste code from your favorite technical books directly into your application.

The perks don't stop there, you can get exclusive access to discounts, newsletters, and great free content in your inbox daily

Follow these simple steps to get the benefits:

1. Scan the QR code or visit the link below

https://packt.link/free-ebook/9781803247878

2. Submit your proof of purchase
3. That's it! We'll send your free PDF and other benefits to your email directly

Part 1:
The Big Picture

This part begins our journey into the world of **Test-Driven Development** (TDD) and provides us with all the essentials we need to start using it for unit testing our code. We begin with an introduction to the principles and practices of TDD, including how it fits into Agile development. Then, we focus our attention on how to apply these practices to write Go unit tests, exploring the fundamentals of test writing and running in Go. Based on these essentials, we explore how to write isolated tests with mocks and simplify our assertions. We learn how to use the third-party open source assertion libraries, `ginkgo` and `testify`, which supplement Go's standard `testing` library. Finally, we learn how to implement and leverage the popular technique of table-driven testing to easily write tests that cover a variety of scenarios, making it easy to cover multiple test scenarios and extend the scope of our tests. In this section, we also begin the implementation of our demo REST API, the `BookSwap` web application.

This part has the following chapters:

- *Chapter 1, Getting to Grips with Test-Driven Development*
- *Chapter 2, Unit Testing Essentials*
- *Chapter 3, Mocking and Assertion Frameworks*
- *Chapter 4, Building Efficient Test Suites*

1

Getting to Grips with
Test-Driven Development

Programs and software have never been more complex than they are today. From my experience, the typical tech startup setup involves deployment to the cloud, distributed databases, and a variety of software integrations from the very beginning of the project. As we use software and consume data at unprecedented rates, the expectation of high availability and scalability has become standard for all the services we interact with.

So, why should we care about testing when we are so busy delivering complex functionality in fast-paced, high-growth environments? Simply put, to verify and prove that the code you write behaves and performs to the expectations and requirements of your project. This is important to you as the software professional, as well as to your team and product manager.

In this chapter, we'll look at the **Agile** technique of **Test-Driven Development** (TDD) and how we can use it to verify production code. TDD puts test writing before implementation, ensuring that test scripts cover and change with requirements. Its techniques allow us to deliver quality, well-tested, and maintainable code. The task of software testing is a necessity for all programmers, and TDD seamlessly incorporates test writing into the code delivery process.

This chapter begins our exploration into the world of testing. It will give you the required understanding of TDD and its main techniques. Defining and setting these fundamentals firmly in our minds will set the stage for the later implementation of automated testing in Go.

In this chapter, we'll cover the following main topics:

- The world and fundamentals of TDD
- The benefits and use of TDD
- Alternatives to TDD
- Test metrics

Exploring the world of TDD

In a nutshell, TDD is a technique that allows us to write automated tests with short feedback loops. It is an iterative process that incorporates testing into the software development process, allowing developers to use the same techniques for writing their tests as they use for writing production code.

TDD was created as an Agile working practice, as it allows teams to deliver code in an iterative process, consisting of writing functional code, verifying new code with tests, and iteratively refactoring new code, if required.

Introduction to the Agile methodology

This precursor to the Agile movement was the **waterfall methodology**, which was the most popular project management technique. This process involves delivering software projects in stages, with work starting on each stage once the stage before it is completed, just like water flows downstream. *Figure 1.1* shows the five stages of the waterfall methodology:

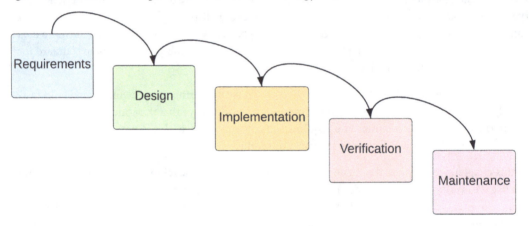

Figure 1.1 – The five stages of the waterfall methodology

Intuition from manufacturing and construction projects might suggest that it is natural to divide the software delivery process into sequential phases, gathering and formulating all requirements at the start of the project. However, this way of working poses three difficulties when used to deliver large software projects:

- Changing the course of the project or requirements is difficult. A working solution is only available at the end of the process, requiring verification of a large deliverable. Testing an entire project is much more difficult than testing small deliverables.

- Customers need to decide all of their requirements in detail at the beginning of the project. The waterfall allows for minimal customer involvement, as they are only consulted in the requirements and verification phases.

- The process requires detailed documentation, which specifies both requirements and the software design approach. Crucially, the project documentation includes timelines and estimates that the clients need to approve prior to project initiation.

> **The waterfall model is all about planning work**
>
> Project management with the waterfall methodology allows you to plan your project in well-defined, linear phases. This approach is intuitive and suitable for projects with clearly defined goals and boundaries. In practice, however, the waterfall model lacks the flexibility and iterative approach required for delivering complex software projects.

A better way of working named **Agile** emerged, which could address the challenges of the waterfall methodology. TDD relies on the principles of the Agile methodology. The literature on Agile working practices is extensive, so we won't be looking at Agile in detail, but a brief understanding of the origins of TDD will allow us to understand its approach and get into its mindset.

Agile software development is an umbrella term for multiple code delivery and project planning practices such as **SCRUM**, **Kanban**, **Extreme Programming** (**XP**), and TDD.

As implied by its name, it is all about the ability to respond and adapt to change. One of the main disadvantages of the waterfall way of working was its inflexibility, and Agile was designed to address this issue.

The **Agile manifesto** was written and signed by 17 software engineering leaders and pioneers in 2001. It outlines the 4 core values and 12 central principles of Agile. The manifesto is available freely at agilemanifesto.org.

The four core Agile values highlight the spirit of the movement:

- **Individuals and interactions over processes and tools**: This means that the team involved in the delivery of the project is more important than their technical tools and processes.

- **Working software over comprehensive documentation**: This means that delivering working functionality to customers is the number one priority. While documentation is important, teams should always focus on consistently delivering value.

- **Customer collaboration over contract negotiation**: This means that customers should be involved in a feedback loop over the lifetime of the project, ensuring that the project and work continue to deliver value and satisfy their needs and requirements.

- **Responding to change over following a plan**: This means that teams should be responsive to change over following a predefined plan or roadmap. The team should be able to pivot and change direction whenever required.

Agile is all about people

The Agile methodology is not a prescriptive list of practices. It is all about teams working together to overcome uncertainty and change during the life cycle of a project. Agile teams are interdisciplinary, consisting of engineers, software testing professionals, product managers, and more. This ensures that the team members with a variety of skills collaborate to deliver the software project as a whole.

Unlike the waterfall model, the stages of the Agile software delivery methodology repeat, focusing on delivering software in small increments or iterations, as opposed to the big deliverables of waterfall. In Agile nomenclature, these iterations are called **sprints**.

Figure 1.2 depicts the stages of Agile project delivery:

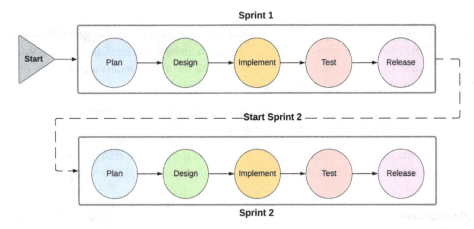

Figure 1.2 – The stages of Agile software delivery

Let's look at the cyclical stages of Agile software delivery:

1. We begin with the **Plan** phase. The product owner discusses project requirements that will be delivered in the current sprint with key stakeholders. The outcome of this phase is the prioritized list of client requirements that will be implemented in this sprint.

2. Once the requirements and scope of the project are settled, the **Design** phase begins. This phase involves both technical architecture design, as well as UI/UX design. This phase builds on the requirements from the **Plan** phase.

3. Next, the **Implement** phase begins. The designs are used as the guide from which we implement the scoped functionality. Since the sprint is short, if any discrepancies are found during implementation, then the team can easily move to earlier phases.

4. As soon as a deliverable is complete, the **Test** phase begins. The **Test** phase runs almost concurrently with the **Implement** phase, as test specifications can be written as soon as the **Design** phase is

completed. A deliverable cannot be considered finished until its tests have passed. Work can move back and forth between the **Implement** and **Test** phases, as the engineers fix any identified defects.

5. Finally, once all testing and implementation are completed successfully, the **Release** phase begins. This phase completes any client-facing documentation or release notes. At the end of this phase, the sprint is considered closed. A new sprint can begin, following the same cycle.

The customer gets a new deliverable at the end of each sprint, enabling them to see whether the product still suits their requirements and inform changes for future sprints. The deliverable of each sprint is tested before it is released, ensuring that later sprints do not break existing functionality and deliver new functionality. The scope and effort of the testing performed are limited to exercising the functionality developed during the sprint.

One of the signatories of the Agile manifesto was software engineer Kent Beck. He is credited with having rediscovered and formalized the methodology of TDD.

Since then, Agile has been highly successful for many teams, becoming an industry standard because it enables them to verify functionality as it is being delivered. It combines testing with software delivery and refactoring, removing the separation between the code writing and testing process, and shortening the feedback loop between the engineering team and the customer requirements. This shorter loop is the principle that gives flexibility to Agile.

We will focus on learning how to leverage its process and techniques in our own **Go** projects throughout the chapters of this book.

Types of automated tests

Automated testing suites are tests that involve tools and frameworks to verify the behavior of software systems. They provide a repeatable way of performing the verification of system requirements. They are the norm for Agile teams, who must test their systems after each sprint and release to ensure that new functionality is shipped without disrupting old/existing functionality.

All automated tests define their inputs and expected outputs according to the requirements of the system under test. We will divide them into several types of tests according to three criteria:

- The amount of knowledge they have of the system
- The type of requirement they verify
- The scope of the functionality they cover

Each test we will study will be described according to these three traits.

System knowledge

As you can see in *Figure 1.3*, automated tests can be divided into three categories according to how much internal knowledge they have of the system they test:

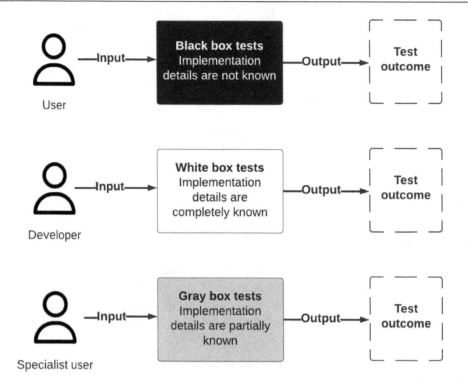

Figure 1.3 – Types of tests according to system knowledge

Let's explore the three categories of tests further:

- **Black box tests** are run from the perspective of the user. The internals of the system are treated as unknown by the test writer, as they would be to a user. Tests and expected outputs are formulated according to the requirement they verify. Black box tests tend not to be brittle if the internals of the system change.

- **White box tests** are run from the perspective of the developer. The internals of the system are fully known to the test writer, most likely a developer. These tests can be more detailed and potentially uncover hidden errors that black box testing cannot discover. White box tests are often brittle if the internals of the system change.

- **Gray box tests** are a mixture of black box and white box tests. The internals of the system are partially known to the test writer, as they would be to a specialist or privileged user. These tests can verify more advanced use cases and requirements than black box tests (for example security or certain non-functional requirements) and are usually more time-consuming to write and run as well.

Requirement types

In general, we should provide tests that verify both the functionality and usability of a system.

For example, we could have all the correct functionality on a page, but if it takes 5+ seconds to load, users will abandon it. In this case, the system is functional, but it does not satisfy your customers' needs.

We can further divide our automated tests into two categories, based on the type of requirement that they verify:

- **Functional tests**: These tests cover the functionality of the system under test added during the sprint, with functional tests from prior sprints ensuring that there are no regressions in functionality in later sprints. These kinds of tests are usually black box tests, as these tests should be written and run according to the functionality that a typical user has access to.

- **Non-functional tests**: These tests cover all the aspects of the system that are not covered by functional requirements but affect the user experience and functioning of the system. These tests cover aspects such as performance, usability, and security aspects. These kinds of tests are usually white-box tests, as they usually need to be formulated according to implementation details.

> **Correctness and usability testing**
>
> Tests that verify the correctness of the system are known as **functional tests**, while tests that verify the usability and performance of the system are known as **non-functional tests**. Common non-functional tests are performance tests, load tests, and security tests.

The testing pyramid

An important concept of testing in Agile is **the testing pyramid**. It lays out the types of automated tests that should be included in the **automated testing suites** of software systems. It provides guidance on the sequence and priority of each type of test to perform in order to ensure that new functionality is shipped with a proportionate amount of testing effort and without disrupting old/existing functionality.

Figure 1.4 presents the testing pyramid with its three types of tests: unit tests, integration tests, and end-to-end tests:

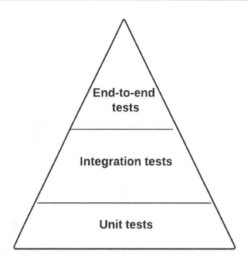

Figure 1.4 – The testing pyramid and its components

Each type of test can then be further described according to the three established traits of system knowledge, requirement type, and testing scope.

Unit tests

At the bottom of the testing pyramid, we have **unit tests**. They are presented at the bottom because they are the most numerous. They have a small testing scope, covering the functionality of individual components under a variety of conditions. Good unit tests should be tested in isolation from other components so that we can fully control the test environment and setup.

Since the number of unit tests increases as new features are added to the code, they need to be robust and fast to execute. Typically, test suites are run with each code change, so they need to provide feedback to engineers quickly.

Unit tests have been traditionally thought of as **white-box tests** since they are typically written by developers who know all the implementation details of the component. However, Go unit tests usually only test the exported/public functionality of the package. This brings them closer to **gray-box tests**.

We will explore unit tests further in *Chapter 2, Unit Testing Essentials*.

Integration tests

In the middle of the testing pyramid, we have **integration tests**. They are an essential part of the pyramid, but they should not be as numerous and should not be run as often as **unit tests**, which are at the bottom of the pyramid.

Unit tests verify that a single piece of functionality is working correctly, while integration tests extend the scope and test the communication between multiple components. These components can be external or internal to the system – a database, an external API, or another microservice in the system. Often, integration tests run in dedicated environments, which allows us to separate production and test data as well as reduce costs.

Integration tests could be **black-box tests** or **gray-box tests**. If the tests cover external APIs and customer-facing functionality, they can be categorized as **black-box tests**, while more specialized security or performance tests would be considered **gray-box tests**.

We will explore integration tests further in *Chapter 4, Building Efficient Test Suites*.

End-to-end tests

At the top of the testing pyramid, we have **end-to-end tests**. They are the least numerous of all the tests we have seen so far. They test the entire functionality of the application (as added during each sprint), ensuring that the project deliverables are working according to requirements and can potentially be shipped at the conclusion of a given sprint.

These tests can be the most time-consuming to write, maintain, and run since they can involve a large variety of scenarios. Just like integration tests, they are also typically run in dedicated environments that mimic production environments.

There are a lot of similarities between integration tests and end-to-end tests, especially in microservice architectures where one service's end-to-end functionality involves integration with another service's end-to-end functionality.

We will explore end-to-end tests further in *Chapter 5, Performing Integration Testing*, and *Chapter 8, Testing Microservice Architectures*.

Now that we understand the different types of automated tests, it's time to look at how we can leverage the Agile practice of TDD to implement them alongside our code. TDD will help us write well-tested code that delivers all the components of the testing pyramid.

The iterative approach of TDD

As we've mentioned before, TDD is an Agile practice that will be the focus of our exploration. The principle of TDD is simple: write the unit tests for a piece of functionality before implementing it.

TDD brings the testing process together with the implementation process, ensuring that every piece of functionality is tested as soon as it is written, making the software development process iterative, and giving developers quick feedback.

Figure 1.5 demonstrates the steps of the TDD process, known as the **red**, **green**, and **refactor** process:

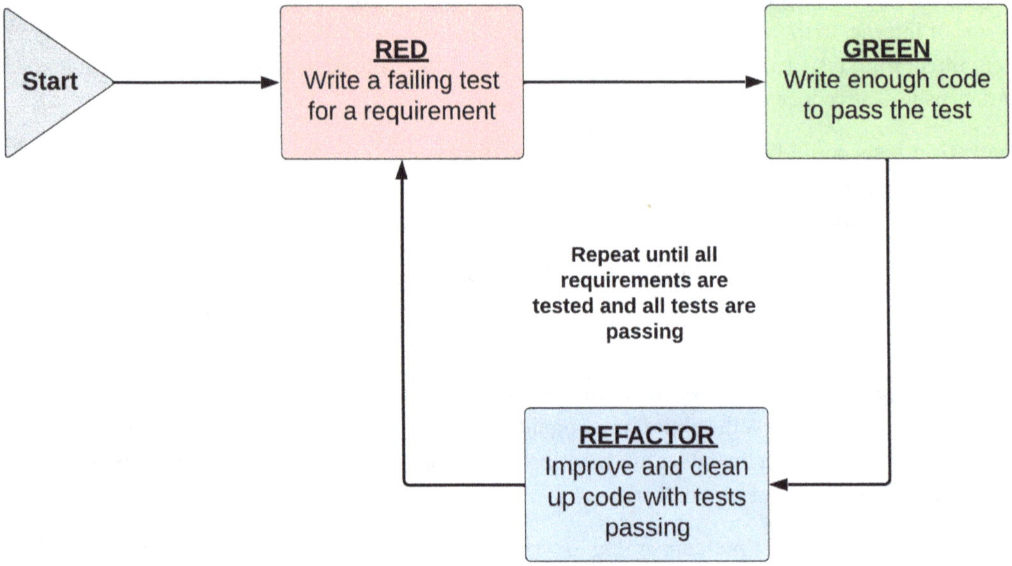

Figure 1.5 – The steps of TDD

Let's have a look at the cyclical phases of the TDD working process:

1. We start at the **red phase**. We begin by considering what we want to test and translating this requirement into a test. Some requirements may be made up of several smaller requirements: at this point, we test only the first small requirement. This test will fail until the new functionality is implemented, giving a name to the red phase. The failing test is key because we want to ensure that the test will fail reliably regardless of what code we write.

2. Next, we move to the **green phase**. We swap from test code to implementation, writing just enough code as required to make the failing test pass. The code does not need to be perfect or optimal, but it should be correct enough for the test to pass. It should focus on the requirement tested by the previously written failing test.

3. Finally, we move to the **refactor phase**. This phase is all about cleaning up both the implementation and the test code, removing duplication, and optimizing our solution.

4. We repeat this process until all the requirements are tested and implemented and all tests pass. The developer frequently swaps between testing and implementing code, extending functionality and tests accordingly.

That's all there is to doing TDD!

TDD is all about developers

TDD is a developer-centric process where unit tests are written before implementation. Developers first write a failing test. Then, they write the simplest implementation to make the test pass. Finally, once the functionality is implemented and working as expected, they can refactor the code and test as needed. The process is repeated as many times as necessary. No piece of code or functionality is written without corresponding tests.

TDD best practices

The **red**, **green**, and **refactor** approach to TDD is simple, yet very powerful. While the process is simple, we can make some recommendations and best practices for how to write components and tests that can more easily be delivered with TDD.

Structure your tests

We can formulate a shared, repeatable, test structure to make tests more readable and maintainable. *Figure 1.6* depicts the **Arrange-Act-Assert (AAA)** pattern that is often used with TDD:

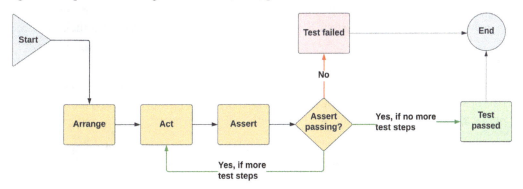

Figure 1.6 – The steps of the Arrange-Act-Assert pattern

The AAA pattern describes how to structure tests in a uniform manner:

1. We begin with the **Arrange** step, which is the setup part of the test. This is when we set up the **Unit Under Test** (**UUT**) and all of the dependencies that it requires during setup. We also set up the inputs and the preconditions used by the test scenario in this section.

2. Next, the **Act** step is where we perform the actions specified by the test scenario. Depending on the type of test that we are implementing, this could simply be invoking a function, an external API, or even a database function. This step uses the preconditions and inputs defined in the **Arrange** step.

3. Finally, the **Assert** step is where we confirm that the UUT behaves according to requirements. This step compares the output from the UUT with the expected output, as defined by the requirements.

4. If the **Assert** step shows that the actual output from the UUT is not as expected, then the test is considered failed and the test is finished.

5. If the **Assert** step shows that the actual output from the UUT is as expected, then we have two options: one option is that if there are no more test steps, the test is considered passed and the test is finished. The other option is that if there are more test steps, then we go back to the **Act** step and continue.

6. The **Act** and **Assert** steps can be repeated as many times as necessary for your test scenario. However, you should avoid writing lengthy, complicated tests. This is described further in the best practices throughout this section.

Your team can leverage test helpers and frameworks to minimize setup and assertion code duplication. Using the AAA pattern will help to set the standard for how tests should be written and read, minimizing the cognitive load of new and existing team members and improving the maintainability of the code base.

Control scope

As we have seen, the scope of your test depends on the type of test you are writing. Regardless of the type of test, you should strive to restrict the functionality of your components and the assertions of your tests as much as possible. This is possible with TDD, which allows us to test and implement code at the same time.

Keeping things as simple as possible immediately brings some advantages:

- Easier debugging in the case of failures
- Easier to maintain and adjust tests when the Arrange and Assert steps are simple
- Faster execution time of tests, especially with the ability to run tests in parallel

Test outputs, not implementation

As we have seen from the previous definitions of tests, they are all about defining inputs and expected outputs. As developers who know implementation details, it can be tempting to add assertions that verify the inner workings of the UUT.

However, this is an anti-pattern that results in a tight coupling between the test and the implementation. Once tests are aware of implementation details, they need to be changed together with code changes. Therefore, when structuring tests, it is important to focus on testing externally visible outputs, not implementation details.

Keep tests independent

Tests are typically organized in test suites, which cover a variety of scenarios and requirements. While these test suites allow developers to leverage shared functionality, tests should run independently of each other.

Tests should start from a pre-defined and repeatable starting state that does not change with the number of runs and order of execution. Setup and clean-up code ensures that the starting point and end state of each test is as expected.

It is, therefore, best that tests create their own UUT against which to run modifications and verifications, as opposed to sharing one with other tests. Overall, this will ensure that your test suites are robust and can be run in parallel.

Adopting TDD and its best practices allows Agile teams to deliver well-tested code that is easy to maintain and modify. This is one of many benefits of TDD, which we will continue to explore in the next section.

Understanding the benefits and use of TDD

With the fundamentals and best practices of TDD in mind, let us have a more in-depth look at the benefits of adopting it as practice in your teams. As Agile working practices are industry standard, we will discuss TDD usage in Agile teams going forward. Incorporating TDD in the development process immediately allows developers to write and maintain their tests more easily, enabling them to detect and fix bugs more easily too.

Pros and cons of using TDD

Figure 1.7 depicts some of the pros and cons of using TDD:

Pros	Cons
+ Allows the development and testing processes to be done at once + Developers analyse requirements in the beginning of each sprint + Increased confidence of code changes + Developers have ownership of their code quality	- Requires writing more code up-front - Requirements need to be elaborated in more detail - Test maintenance can be time consuming

Figure 1.7 – Pros and cons of using TDD

We can expand on these pros and cons highlights:

- TDD allows the development and testing process to happen at the same time, ensuring that all code is tested from the beginning. While TDD does require writing more code upfront, the written code is immediately covered by tests, and bugs are fixed while relevant code is fresh in developers' minds. Testing should not be an afterthought and should not be rushed or cut if the implementation is delayed.

- TDD allows developers to analyze project requirements in detail at the beginning of the sprint. While it does require product managers to establish the details of what needs to be built as part of sprint planning, it also allows developers to give early feedback on what can and cannot be implemented during each sprint.

- Well-tested code that has been built with TDD can be confidently shipped and changed. Once a code base has an established test suite, developers can confidently change code, knowing that existing functionality will not be broken because test failures would flag any issues before changes are shipped.

- Finally, the most important pro is that it gives developers ownership of their code quality by making them responsible for both implementation and testing. Writing tests at the same time as code gives developers a short feedback loop on where their code might be faulty, as opposed to shipping a full feature and hearing about where they missed the mark much later.

In my opinion, the most important advantage of using TDD is the increased ownership by developers. The immediate feedback loop allows them to do their best work, while also giving them peace of mind that they have not broken any existing code.

Now that we understand what TDD and its benefits are, let us explore the basic application of TDD to a simple calculator example.

Use case – the simple terminal calculator

This use case will give you a good understanding of the general process we will undertake when testing more advanced examples.

The use case we will look at is the simple terminal calculator. The calculator will run in the terminal and use the standard input to read its parameters. The calculator will only handle two operators and the simple mathematical operations you see in *Figure 1.8*:

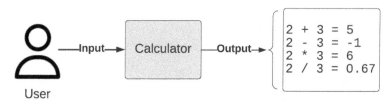

Figure 1.8 – The simple calculator runs in the terminal

This functionality is simple, but the calculator should also be able to handle edge cases and other input errors.

Requirements

Agile teams typically write their requirements from the user's perspective. The requirements of the project are written first in order to capture customer needs and to guide the test cases and implementation of the entire simple calculator project. In Agile teams, requirements go through multiple iterations, with engineering leadership weighing in early to ensure that the required functionality can be delivered.

Users should be able to do the following:

- Input positive, negative, and zero values using the terminal input. These values should be correctly transformed into numbers.
- Access the mathematical operations of addition, subtraction, multiplication, and division. These operations should return the correct results for the entire range of inputs.
- View fractional results rounded up to two decimal places.
- View user-friendly error messages, guiding them on how to fix their input.

> **Agile requirements from the perspective of the user**
>
> Requirements are used to capture the needs and perspectives of the end user. The requirements set out the precondition, the user actions, and the acceptance criteria. They specify what we should build as well as how to verify the implementation.
>
> Remember that we only specify requirements on a sprint-by-sprint basis. It is an anti-pattern to specify requirements of the entire product upfront, as well as work in the mindset that they cannot change. Software building in Agile is an iterative process.

Architecture

Our simple terminal calculator is small enough to implement in one sprint. We will take our four requirements and translate them into a simple system architecture. The calculator will be downloaded and run by users locally, so we do not need to consider any networking or cloud deployment aspects.

Figure 1.9 shows what the design of the calculator module could look like:

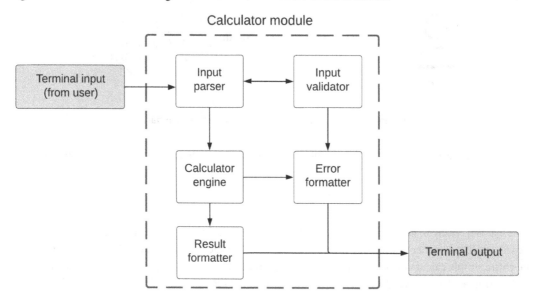

Figure 1.9 – Architecture of the simple terminal calculator

Each of the components of the calculator module has its own, well-defined responsibilities and functionality:

- The **Input parser** is in charge of integrating with the terminal input and reading the user input correctly and passing it to the calculator module.

- The **Input validator** is in charge of validating the input sent from the **Input parser**, such as whether the input contains valid numbers and the operators are valid.

- Once the input is parsed and validated, the **Calculator engine** takes in the numbers and attempts to find the result of the operation.

- The **Calculator engine** then relies on the **Result formatter** to format the result correctly and print it to the terminal output. In the case of an error, it relies on the **Error formatter** to produce and print user-friendly errors.

Applying TDD

As described, we will use the **red**, **green**, and **refactor** process to apply TDD to deliver the required user functionality in an iterative manner. Tests are written first, based on the requirements and design of the simple terminal calculator.

An overview of how the process might work for the implementation of the `Divide (x,y)` function in **the calculator engine** is demonstrated in *Figure 1.10*:

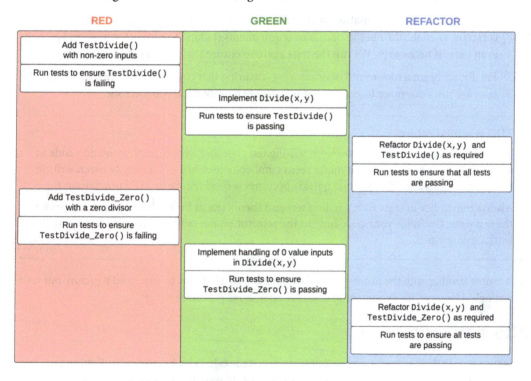

Figure 1.10 – The TDD process applied to the calculator engine

This is a small snapshot that demonstrates the steps involved when using TDD:

1. We begin by writing a simple `TestDivide ()` that arranges two non-zero inputs and writes assertions for dividing them. This is the simplest case that we can implement. Then, we run the test suite to ensure that the newly written `TestDivide ()` is failing.

2. Now that the test has established the expected behavior, we can begin our implementation of the `Divide(x,y)` function. We write just enough code to handle the simple case of two non-zero inputs. Then, we run the test suite to verify that the code we have written satisfies the assertions of `TestDivide()`. All tests should now be passing.

3. We can now take some time to refactor the existing code that we have written. The newly written code can be cleaned up according to the clean code practices, as well as the TDD best practices that we have discussed. The test suite is run once more to validate that the refactor step has not broken any new or existing tests.

4. The simplest functionality for the new `Divide(x,y)` function is now implemented and validated. We can turn to looking at more advanced functionality or edge cases. One such edge case could be handling a zero divisor gracefully. We now add a new test, `TestDivide_Zero()`, which sets up and asserts the case of a zero divisor. As usual, we run the test suite to ensure that the new `TestDivide_Zero()` test is failing.

5. We modify the implementation of `Divide(x,y)` to handle a zero divisor gracefully and correctly, as established in the calculator requirements (talking to product owners and perhaps even users if necessary). We run the tests again to ensure that all tests are now passing.

6. Finally, we begin a new round of refactoring, ensuring that code and tests are well written. All tests are run once more to ensure that refactoring has not caused any errors.

> **TDD is second nature**
>
> The development process swaps between writing test code and writing implementation code as many times as required. While it might seem cumbersome at first, swapping between writing test code and implementation code quickly becomes second nature to TDD practitioners.
>
> Always remember to start with a failing test and then write as little code as possible to make the test pass. Optimize your code only in the refactor phase, once you have all functionality working as verified.

We are now familiar with the process of TDD and have looked at how to write and structure our tests accordingly. However, it's important to consider alternative processes as well.

Alternatives to TDD

As we've seen, TDD is simply a way to deliver well-tested code in an iterative way. Putting tests first ensures that no functionality is ever delivered without being tested and refactored. In this section, we will have a look at some other common processes for testing code.

Waterfall testing

As we remember from our introduction to the waterfall methodology, the testing or verification phase of waterfall projects happens after the implementation phase is fully completed. The entire project is delivered, and all requirements are implemented by this point.

Here are the advantages:

- Waterfall projects are typically well structured and well documented. Testing plans are informed by this extensive documentation and testers can ensure that all of the end-to-end tests that they implement cover the identified user needs.

- Developers and testers can rely on the project documentation to work independently, without the need to communicate. This division allows teams to work in shifts – testers verify functionality and developers fix any bugs that may arise.

These are the disadvantages:

- As the entire project is already implemented, it is easier for bugs to become complex. Furthermore, since the entire project is already implemented, it might take considerably more engineering effort to fix a bug, in the case that large changes need to be undertaken.

- In the case that client requirements are not well known or clear from the beginning, a lot of implementation and testing effort might be wasted if the requirements change once the client sees the delivered product at the end of the process.

- The testing process can often be seen as a time-wasting, negative exercise that should be finished as soon as possible. Furthermore, if there are delays in the development process, it can be easy to cut corners in the verification process, delivering an unstable product.

Acceptance Test-Driven Development

Acceptance Test-Driven Development (ATDD) is an Agile development process related to TDD. ATDD involves people from multiple disciplines from product, engineering, and testing to ensure that **the right product is being developed in the right way**. The customer requirements are translated into a list of requirements that can be understood by a wide variety of stakeholders. These requirements are then converted to automated acceptance tests, which are used to verify what the engineering department is delivering.

The advantages of ATDD are as follows:

- Just like with TDD, tests are written first when you use ATDD. A complete suite of automated acceptance tests can be run after each commit or incremental code delivery, ensuring that all end-to-end functionality works as expected.

- If done right, using ATDD on a project will be widely supported by a wide variety of stakeholders inside the business, as they will have a good understanding of the direction and customer value it will provide.

The disadvantages are as follows:

- Significant communication and synchronization effort is required for the inter-disciplinary effort of writing requirements. It can be time-consuming to get a variety of stakeholders to give the time and effort needed.

- This approach might not be best suited for greenfield projects, where there are a lot of unknowns upfront. It can be particularly challenging to write acceptance tests for a project that does not even have an API or database model yet.

- It can be challenging to get sample payloads or datasets from the outset of a project, especially if these are provided by the client or a third party.

Further related to ATDD, we have **Behavior-Driven Development** (**BDD**). It provides precise guidance on how to structure the conversation between stakeholders using business domain language. We will explore BDD further in *Chapter 5, Performing Integration Testing*.

As we begin to write and think of test code together with functional code, it's important to set success criteria for our test code. Test metrics can help us achieve just that.

Understanding test metrics

Now that we understand how to deliver projects with tests first, it's time to look at some metrics that can quantify how well-tested a project is. It's important to deliver tests across the entire test pyramid, as it's important to be able to ensure the application is working correctly end-to-end as well as working well with its external dependencies.

Important test metrics

There is a wide range of metrics that we can measure when quantifying the quality of software:

- **Requirement coverage**: This indicates the percentage of your project requirements that are covered by tests. A test could cover multiple requirements, but no customer requirement should be left untested.

- **Defect count and distribution**: This indicates how many defects or bugs are discovered in each module/part of the application. The distribution will also signal whether there are any particular problem areas in the system that could be refactored.

- **Defect resolution time**: This indicates how quickly the development team is able to fix bugs once they are detected. A long **Mean Time To Resolution** (**MTTR**) can indicate that the development team is short-staffed, while a long max resolution time in a particular area of the system can indicate that the code in that particular part is difficult to change.

- **Code coverage**: This indicates the percentage of your code base that is exercised by unit tests. Since tests should be written first, coverage also shows whether the development team is using TDD. Low test coverage can also indicate issues with the system design.

- **Burndown rates and charts**: These indicate the rate at which the team is able to deliver functionality. As development and testing are a unified task, a user story or requirement cannot be considered complete unless it is tested, so the burndown rate will include only stories that are ready for delivery. Burndown charts can indicate delays in project timelines.

Code coverage

Since the code coverage metric is such an important TDD indicator, let's explore it further. In order to achieve a high coverage percentage, tests should cover the following:

- The **functions** you implemented

- The **statements** that your functions are composed of

- The different **execution paths** of your functions

- The different **conditions** of your Boolean variables

- The different **parameter values** that can be passed to your functions

The Go test runner provides the coverage percentage for Go applications. We will have a look at how to do this in *Chapter 2, Unit Testing Essentials*.

Figure 1.11 shows a flow chart of the implementation of the Divide (x,y) function from the simple terminal calculator:

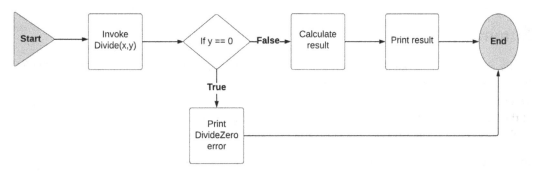

Figure 1.11 – Execution flow of the Divide function in the simple calculator

Tests should be written to cover and verify the following:

- The execution path for `y != 0`
- The execution path for `y == 0`
- The error message of the `DivideZero` error
- The output from the result calculation statements
- The output from the print result statements

Code coverage percentage

In large projects, it will be unfeasible to reach 100% test coverage for the code base. There have been many discussions in the tech community about what a *good* test coverage percentage is. It is generally accepted that a good coverage amount is around the 80% mark. After that point, experience shows there can be diminishing returns.

The code coverage percentage will also depend on the kind of project you are running. A legacy code base with a low code coverage percentage will require considerable effort to bring up to the 80% mark. Similarly, a greenfield project will also be difficult to test if there are many unknowns.

Just like any code you write, test code needs to be maintained and refactored. Keeping a high coverage percentage requires maintaining and updating a lot of test code. This can increase the development cost of refactoring or other code changes, as it potentially requires updating many test cases. The business value of maintaining tests that do not cover requirements is very low, so you should ensure that your tests are providing value to your test suites.

> **Well-tested code is not necessarily bug-free code**
>
> Your tests should aim to provide verification for important code behavior, as opposed to simply writing code to get a certain code coverage percentage. The team should embrace a testing culture using TDD and a good coverage percentage will follow.

Summary

In this chapter, we covered all the testing fundamentals that you will need to get started with TDD. We began with an explanation of what Agile is and how TDD fits into the Agile development process. You learned about the different types of automated tests and the testing pyramid. Then, we looked at the iterative process of delivering code with TDD using the red, green, and refactor process, and explored some TDD best practices on how to structure and write tests.

In *Chapter 2, Unit Testing Essentials*, we will learn how to write tests in Go and begin to get some hands-on experience with TDD. We will begin to use the red, green, and refactor process and write tests according to the TDD best practices that we have learned.

Questions

1. What is the testing pyramid? What are its components?

2. What is the difference between functional and non-functional tests?

3. Explain what the red, green, and refactor TDD approach is.

4. What is ATDD?

5. What is code coverage?

Further reading

*   *Learning Agile: Understanding Scrum, XP, Lean, and Kanban* – Andrew Stellman and Jennifer Greene, published by O'Reilly Media

*   *Test Driven Development: By Example* – Kent Beck, published by Addison-Wesley Signature Series

*   *Clean Code: A Handbook of Agile Software Craftsmanship* – Robert C. Martin, published by Prentice Hall

Answers

1. The testing pyramid specifies how automated test suites should be structured. At the bottom of the pyramid are unit tests, which test a single isolated component. Next up in the middle of the pyramid are integration tests, which test that multiple components are able to work together as specified. Finally, at the top of the test pyramid are end-to-end tests that test the behavior of the entire application.

2. Functional tests cover the correctness of a system, while non-functional tests cover the usability and performance of a system. Both types of tests are required to ensure that the system satisfies the customers' needs.

3. The red, green, and refactor TDD approach refers to the three phases of the process. The red phase involves writing a new failing test for the functionality we intend to implement. The green phase involves writing enough implementation code to make all tests pass. Finally, the refactor phase involves optimizing both implementation and testing code to remove duplication and come up with better solutions.

4. Acceptance test-driven development. Just like TDD, ATDD puts tests first. ATDD is related to TDD, but it involves writing a suite of acceptance tests before the implementation begins. It involves multiple stakeholders to ensure that the acceptance test captures the customer's requirements.

5. Code coverage is the percentage of your lines of code that are exercised by your unit test. This is calculated by considering the function statements, parameter values, and execution paths of your code. The Go test runner outputs the calculated code coverage. We should aim for a good value, but optimizing for 100% is normally not appropriate.

2
Unit Testing Essentials

In the previous chapter, we learned all about the iterative process of writing tests alongside code using TDD, as well as how it fits into the **Agile project methodology**. We covered the **red, green, refactor approach**, which requires frequent switching between source code and test code.

When first starting with TDD, following a prescribed process for writing code may seem like an artificial way of working, but it soon becomes second nature with practice. In this chapter, we will learn all the Go fundamentals required to begin putting everything else we have learned to use. We will begin to write unit tests with Go's testing package, focusing on the test writing syntax and process. This chapter will allow us to get hands-on experience with all the concepts we have explored so far.

As we saw from the **testing pyramid** in *Chapter 1, Getting to Grips with Test-Driven Development*, **unit tests** are the most numerous. They are used to verify the functionality of a single unit, in isolation. We will begin our exploration of TDD and Go testing by implementing unit tests.

The Go programming language was created by the team at Google to allow developers to write simple and efficient software, which they felt was not achievable with the tools they had available at the time. This principle of simplicity is echoed throughout the entire language, including its test writing and running functionality.

The Go standard library provides the testing package, which provides the essentials we need for writing automated tests. Tests are simply functions that satisfy certain conventions and signatures, making it easier for developers to write code using the same strategies and mechanisms as they write their source code.

The Go toolchain provides a single go test command for running all the tests that we have defined. This simple yet powerful command also provides functionality for running benchmarks, which we can use to examine the performance of a given component.

In this chapter, we will cover the following main topics:

- The Go package as the **Unit Under Test (UUT)**

- The fundamentals of working with Go's testing package

- Test setup and teardown

- Grouping tests into suites with subtests

- The difference between a test and a benchmark

Technical requirements

You will need to have **Go version 1.19** or later installed to run the code samples in this chapter. The installation process is described in the official Go documentation at `https://go.dev/doc/install`.

The code examples included in this book are publicly available at `https://github.com/PacktPublishing/Test-Driven-Development-in-Go/chapter02`.

The unit under test

In *Chapter 1, Getting to Grips with Test-Driven Development*, we discussed the structure of tests using the **Arrange-Act-Assert (AAA)** pattern. We also briefly mentioned that the Arrange step sets up the **Unit Under Test (UUT)** and its dependencies. The test then exercises and verifies the functionality of the UUT.

In Go, source code is organized into packages and modules. We will begin by exploring what these are and how they work and then look at how test files fit into this structure. A solid understanding of the power of packages will set the scene for us to begin considering not only *how* to write tests, but *what* to test.

Modules and packages

If you have worked with Go for a while, you might be familiar with Go's **module system**, which was introduced as the default dependency management solution in **Go 1.13**. The latest Go version at the time of writing is **version 1.19**, so the module system has been the standard solution for some time.

Modules

A **module** is a collection of packages that are distributed and released together. The name of the module should be representative of its functionality, as well as where to find it. The Go toolchain provides support for popular code hosting solutions such as GitHub and Bitbucket and can issue the correct requests for downloading dependencies from their version control systems.

Modules are declared by placing a go.mod file at the root of the project directory. A new module can be initialized by running the go mod init command with a module path as a parameter. We can initialize the module for all our code examples by running the following command in the Terminal:

```
$ go mod init "github.com/PacktPublishing/Test-Driven-
Development-in-Go"
```

The module path we have provided is the same as the GitHub repository path. The generated file only contains two lines – one for the module path and one for the version of Go required:

```
module github.com/PacktPublishing/Test-Driven-Development-in-Go
go 1.19
```

In general, the go.mod file contains these properties:

- The module path
- The version of Go required for the project
- Any external dependencies that need to be imported when building the project

Since our project is empty for now, our module's file does not specify any external dependencies. The packages of the standard library do not need to be declared as dependencies. This file will be automatically updated with dependencies once they are used in the source code of the module.

> **The generated go.mod file**
>
> The go.mod file is generated, but not read-only. However, as a rule of thumb, you should avoid editing the go.mod file manually. In general, developers edit it only to change version numbers, not to manually change its entries. You can also recreate it any time with the go mod init command.

Packages

While modules are a fantastic way to bundle and release projects, most production systems would be almost impossible to maintain or understand if they did not have any internal organization or hierarchy. This is where Go packages come in to help us provide this much-needed structure.

Go source code is organized into **packages**. The first line of every source file must be the package declaration, which can be done using the package keyword. All the names defined in the source file are then added to the declared package.

Looking back at the simple terminal calculator example from *Chapter 1, Getting to Grips with Test-Driven Development*, we can specify a package and source code structure for it, as seen in *Figure 2.1*:

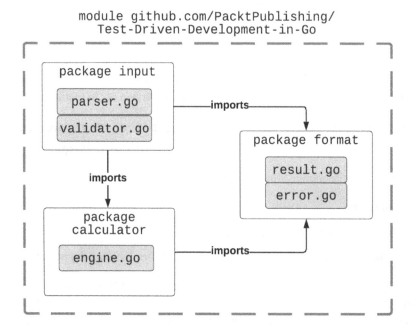

Figure 2.1 – The module, packages, and source files of the simple Terminal calculator

The module contains three packages, each containing specialized functionality:

- The input package contains an input parser and validation functionality. It has dependencies on the calculator and format packages.

- The calculator package contains all the calculation engine logic, providing functionality for all the operations that the calculator provides. It has a dependency on the format package.

- The format package contains formatting logic for results and errors. It has no dependencies on our existing packages.

Package naming is important

Package names should be representative of the functionality they provide so that other code may reference them. Package names should be short and descriptive. They should also be meaningful when used along with the names of the types and functions they provide.

The `format` package is at the bottom of our package hierarchy, and we can begin by defining it and its result formatting capabilities immediately. Looking at the contents of its `result.go` source file, we can see its simple definition:

```
package format

func Result(expression string, result float64) string {
    // implementation code
    return ""
}
```

This package defines a `Result` function that outputs a formatted string of the given expression and result. We added the return of an empty string to ensure the code compiles until we are ready to begin implementation with TDD. The `error.go` file is similarly defined and has been omitted for brevity.

Looking at the calculator engine in more detail, the contents of the `engine.go` source file can begin like so:

```
package calculator

type Engine struct {}

func(e *Engine) Add(x, y float64) float64{
    // implementation code
    return 0
}

// ... method declarations
```

We begin with the `package calculator` definition, adding the source file and all its definitions to the `calculator` package. Then, we create an `Engine` type that will contain all the dependencies of the calculator. After these few lines, we can begin to define methods for all the operations that the engine needs to provide. The `Add` method of the `Engine` type is an example of what the definition of the addition operation is.

Those of you who are eagle-eyed will notice that the types, methods, and functions have been defined with a capital letter. This makes them **exported names**.

> **Visibility outside of the package**
>
> Only the exported names of a package are visible for usage outside of their defined package. Unlike other languages, there are no visibility modifiers in Go. In the code examples for the `format` and `calculator` packages, we need their functionality to be available outside of their respective packages, so this is why they have been exported and defined with a capital letter.

A package can declare a dependency on another package by using the `import` keyword. We can then reference the variables, types, and functions of the imported package by qualifying them with the package name and the **dot operator** (.).

With some exceptions, there can only be one single package name per directory. The declaration of the new `input` package will also require the creation of a new directory. The declaration of `Parser`, which has a dependency on the `Engine` type, looks like this:

```
package input

import "github.com/PacktPublishing/Test-Driven-Development-in-
Go/chapter02/calculator"

type Parser struct {
    engine *calculator.Engine
    validator *Validator
}

func (p *Parser) ProcessExpression(expr string) (*string,
error) {
    // implementation code
}

// ... method declarations
```

`Parser` is declared as part of `package input`. The `Parser` type needs the functionality of the `Calculator` type, so it imports the `calculator` package. As mentioned, the reference to `calculator.Engine` is qualified using its provided package and the dot operator. This lets the compiler know that the type referenced is coming from an imported package, not the current package.

The highlighted import path consists of three parts:

- The module that the package belongs to: `github.com/PacktPublishing/Test-Driven-Development-in-Go`

- The sub-directory path from the root of the module: `chapter02`
- The name of the imported package: `calculator`

The power of Go packages

Packages are a powerful and central concept to Go. They allow developers to do the following:

- **Group components**: When named well, packages provide an easy way to understand, uniquely group, and document multiple components that share the same functionality.
- **Encapsulate code**: Since only exported methods are visible to external code, packages are the most important encapsulation mechanism in Go. They give developers fine-grained control over exactly what is available for use outside of the package.
- **Reuse code**: Packages provide modularity to our programs, allowing us to reuse code in multiple places by providing a way for users to import them. The ability to leverage code from outside the current module allows developers to share the same solutions, without reinventing the wheel.
- **Easily manage dependencies**: Go's module system follows **Semantic Versioning (SemVer)**, which uses three primary numbers to manage imported dependencies: the major, the minor, and the patch version. This allows developers to pin dependencies to a certain version, as well as easily know when they need to upgrade to a newer version.

Now that we understand the fundamentals of modules and packages, let's turn our attention toward where testing fits into the codebase and its packages. Notice that the methods of the `Engine` and `Parser` custom types do not have any code implemented yet: this is because TDD is all about writing tests first!

> **Packages as APIs**
>
> Due to their encapsulation and modularity properties, packages enable developers to build and structure their code using similar techniques as when designing external APIs, by choosing the signatures and functions they want to provide to external users.

Test file naming and placement

Unlike other programming languages, test files live alongside the source code in Go. All test files must end with the `_test.go` suffix. Go's test runner scans the codebase for these test files and runs them accordingly. The test runner is part of the Go toolchain and can be invoked using the `go test` command.

Figure 2.2 presents the directory structure of the simple terminal calculator discussed so far:

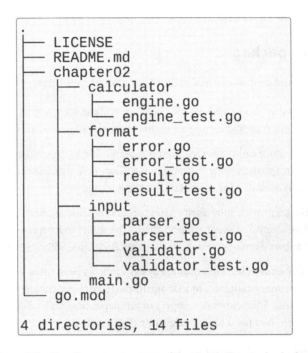

```
.
├── LICENSE
├── README.md
├── chapter02
│       ├── calculator
│       │       ├── engine.go
│       │       └── engine_test.go
│       ├── format
│       │       ├── error.go
│       │       ├── error_test.go
│       │       ├── result.go
│       │       └── result_test.go
│       ├── input
│       │       ├── parser.go
│       │       ├── parser_test.go
│       │       ├── validator.go
│       │       └── validator_test.go
│       └── main.go
└── go.mod

4 directories, 14 files
```

Figure 2.2 – The directory structure of the simple Terminal calculator

All the code that we will be discussing in this chapter can be found in the chapter02 directory of the dedicated repository corresponding to this book. Then, inside this directory, we have three further directories for format, calculator, and input, which each contain their source code files and corresponding test files.

> **Naming test files**
>
> While test files need to end with the _test.go suffix, matching the rest of their name to their corresponding source code file is not enforced. However, it is highly recommended that you use the source filename and then append the test suffix. This will also ensure that the two files stay together when sorted lexicographically.

The source files and test files live directly next to each other, in the same directory, making it even easier for developers to swap between writing implementation code and test code when practicing TDD. Some editing tools can even do this with a keyboard shortcut!

Additional test packages

Although test files are named the same as their corresponding source files and live in the same directory, the package structure will tell a different story. We previously mentioned that, with some exceptions, only one package may be declared per directory. Test files are one of these exceptions.

Test files are allowed to declare an additional test package, matching the source files package with `_test` appended. From a visibility perspective, this test package is the same as any other package and will need to import the packages that it wishes to have access to. It will also only have access to the exported names of its imported packages.

> **Test packages as a recommended practice**
>
> The usage of the dedicated `_test` package is not enforced in Go, but it is recommended. Whenever possible, you should declare your tests in a dedicated test package.

Figure 2.3 depicts the separate definitions of test packages in the simple Terminal calculator:

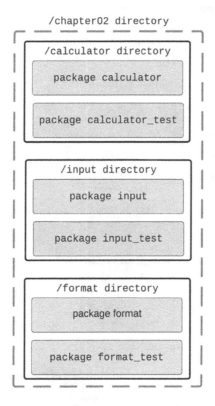

Figure 2.3 – The package and directory structure of the simple calculator

The dedicated test package is defined in the same directory as the source package, achieving full separation between the source and test code. Using the dedicated test package brings the following advantages:

- **Prevents brittle tests**: Restricting access to only exported functionality does not give test code visibility into package internals, such as state variables, which would otherwise cause inconsistent results.

- **Separates test and core package dependencies**: The test package allows the test to import any dependencies required, without adding those dependencies to the core package. In practice, test code will often have its own dedicated verifiers and functionality, which we would not want to be visible to production code. The test package is a seamless way to guarantee separation.

- **Allows developers to integrate with their own packages**: We previously mentioned that packages allow developers to build their internal code as small APIs. Writing tests from a dedicated test package allows developers to see how easy it is to integrate with their designed external interfaces, ensuring that their code is maintainable.

Figure 2.4 shows the updated module, packages, source, and test files of the simple Terminal calculator, which now uses _test packages:

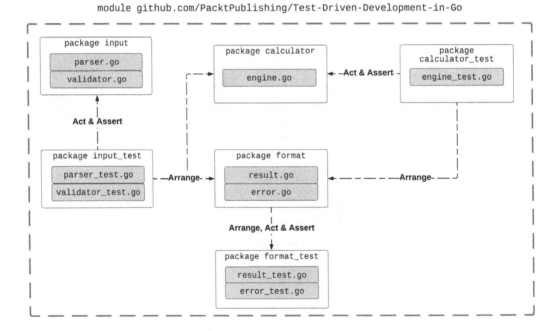

Figure 2.4 – The module, packages, source, and test files of the simple Terminal calculator

Let's describe the dependencies between the packages in terms of the AAA pattern that we know from *Chapter 1, Getting to Grips with Test-Driven Development*. The **Arrange** and **Assert** steps are performed on the UUT and are conveniently defined in the correspondingly named package:

- The `format` package does not have any dependencies on other packages. As a result, the `format_test` package performs all three steps on the `format` package.

- The `calculator` package has a dependency on the `format` package. As a result, the `calculator_test` package arranges the dependencies from the `format` package. Then, it performs the Act and Assert steps on the `calculator` package.

- Finally, the `input` package has a dependency on the `calculator` and input packages. As a result, the `input_test` package arranges the dependencies of the `input` package, which are provided by the `calculator` and `input` packages.

This section has given you an introduction to the Go module system and discussed how to place and name tests in the overall codebase. Now, let's look at how to implement tests in Go.

Working with the testing package

The standard library provides the `testing` package, which contains the essentials we need for writing and running tests. In this section, we will explore how to use it and begin to apply it so that we can write tests for our simple terminal calculator example.

The testing package

The `testing` package provides support for testing Go code. It must be imported by all test code as this is the way to interact with the test runner. At a glance, the `testing` package seems very simplistic, but it fits with Go's language design. Packages should be small, focused, and have a limited number of dependencies. This should make them easy to test with a relatively simple testing library.

Here are some of the important types from the `testing` library that we will be using:

- `testing.T`: All tests must use this type to interact with the test runner. It contains a method for declaring failing tests, skipping tests, and running tests in parallel. We will look at and begin to use these methods in this section.

- `testing.B`: Analogous to the test runner, this type is Go's **benchmark** runner. It has the same methods for failing tests, skipping tests, and running benchmarks in parallel. Benchmarks are special kinds of tests that are used for verifying the performance of your code, as opposed to its functionality. We will explore benchmarks later in this chapter.

- `testing.F`: This type is used to set up and run **fuzz tests** and was added to the Go standard toolchain in **Go 1.18**. It creates a randomized seed for the testing target and works together with the `testing.T` type to provide test-running functionality. Fuzz tests are special kinds of tests that use random inputs to find edge cases and bugs in our code. We will explore fuzz tests further in *Chapter 10, Testing Edge Cases*.

> **The testing package is used in all tests**
>
> The `testing` package must be imported by all tests as it is the only way to interact with Go's test runner. As previously discussed, test filenames must end with the `_test.go` suffix, but tests will only be run if they use the `testing` package. Tests must also satisfy a standard test signature, which is explained in the next section, *Working with the testing package*.

Now, let's have a look at the `testing.T` type in a little bit more detail since it will be the focus of our exploration in this chapter.

Figure 2.5 presents a summary of some of the methods we will discuss:

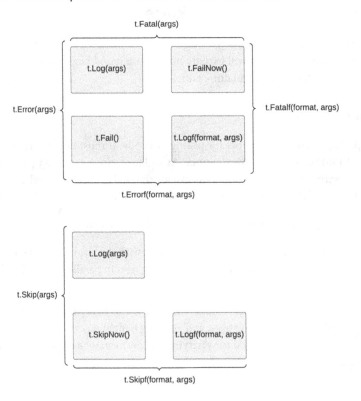

Figure 2.5 – The logging, failing, and skipping methods of the testing.T type

It exposes the following methods for logging, skipping, and failing tests that are important to understand:

- `t.Log(args)`: This prints the given arguments to the error log after the test has finished executing.

- `t.Logf(format, args)`: This has the same functionality as the `t.Log` method, but allows the arguments to be formatted before printing.

- `t.Fail()`: This marks the current test as failed but continues execution until the end.

- `t.FailNow()`: This marks the current test as failed and immediately stops the execution of the current test. The next test will be run while continuing the suite.

- `t.Error(args)`: This is equivalent to calling `t.Log(args)` and `t.Fail()`. This method makes it convenient to log an error to the error log and mark the current test as failed.

- `t.Errorf(format, args)`: This is equivalent to calling `t.Logf(format, args)` and `t.Fail()`. This method makes it convenient to fail a test, then format and print an error line in one call.

- `t.Fatal(args)`: This is equivalent to calling `t.Log(args)` and `t.FailNow()`. This method makes it convenient to fail a test and print an error line in one call.

- `t.Fatalf(format, args)`: This is equivalent to calling `t.Logf(format, args)` and `t.FailNow()`. This method makes it convenient to fail a test, then format and print an error line in one method call.

- `t.SkipNow()`: This marks the current test as skipped and immediately stops its execution. Note that if the test has already been marked as failed, then it remains failed, not skipped.

- `t.Skip(args)`: This is equivalent to calling `t.Log(args)`, followed by `t.SkipNow()`. This method makes it convenient to skip a test and print an error line in one call.

- `T.Skipf(format, args)`: This is equivalent to calling `t.Logf(format, args)`, followed by `t.SkipNow()`. This method makes it convenient to skip a test, then format and print an error line in one call.

In general, developers use the convenience methods presented previously when writing their tests, as opposed to invoking `t.Fail()`, `t.FailNow()`, or `t.SkipNow()` explicitly. Going forward, we will make use of them as we begin to write our test code.

Another thing you might be wondering about is whether the `testing` package provides any assertion functionality. It does not provide any inbuilt assertions, so we will need to compare values ourselves. We will explore third-party assertion libraries further in *Chapter 3, Mocking and Assertion Frameworks*.

Test signatures

The `testing` package is used for writing unit tests, which are placed in their own test files. Go tests are functions that satisfy the following signature:

```
func TestName(t *testing.T) {
  // implementation
}
```

This test signature highlights the following requirements for Go tests:

- Tests are exported functions whose name begins with `Test`.

- Test names can have an additional suffix that specifies what the test is covering. The suffix must also begin with a capital letter, as we can see specified by `Name` in the test signature, which doubles as the test name.

- Tests must take in a single parameter of the `*testing.T` type. As we've explained so far, this will be how the test interacts with the test runner. You can name the testing parameter however you want, but Go developers typically use `t` to denote it.

- Tests must not have a return type.

> **Go tests are just functions**
>
> As we can see, tests are simply functions that satisfy a certain signature. The Go test tool scans the code base for these special functions in the `_test.go` files and runs them accordingly.

Inside these test functions, we can define and implement our test code using the AAA pattern. You should keep the scope of your test small, preferring to write several tests rather than writing one large, potentially brittle test.

Just like package names, test names are very important, so we need to give them some special consideration. Having effectively named tests can bring developers some important advantages:

- **Documentation and understanding**: A suite of effectively named tests will help newcomers understand how a particular piece of code is supposed to work. As they are easy to change, they also allow you to explore the behavior of the code under a variety of conditions.

- **Refactoring support**: The test name sets the intention of the test; then, its implementation simply executes it. Once the code has been refactored, the test implementation may change, but the intention of the test, conveyed by its name, remains. Well-named tests can support code refactoring, which might need to change the test implementation/execution. We will discuss code refactoring strategies further in *Chapter 7, Refactoring in Go*.

- **Consistency**: Setting a standard for how tests should be named and structured throughout your code base will make it easier for you to know what to expect, reducing cognitive load when reading code.

Other than the special signature we've just seen, Go does not enforce any other naming standards. The consensus in the Go community is that they should be easily understood and concise. The Go standard library ties tests with the name of the function that they test: the UUT. The test name simply follows the `TestUnitUnderTest` structure. For example, a test for the `Add` function will be named `TestAdd`.

Another common approach is to name the tests using a **Behavior-Driven Development** (**BDD**) style approach. We will explore BDD tests in detail in *Chapter 5, Performing Integration Testing*.

In this naming approach, the name of the test follows the structure of `TestUnitUnderTest_PreconditionsOrInputs_ExpectedOutput`. For example, a test for the function will be named `TestAdd_TwoNegativeNumbers_NegativeResults` if it tests adding two negative numbers together.

While the BDD style naming pattern is a lot more precise, it goes against the principle of simplicity and conciseness that is so central to Go. We will use the simpler approach: naming the test after its UUT. We will see how we can achieve the extra precision of preconditions and expected output using subtests later in this chapter.

Running tests

One of the commands of the Go toolchain is the `go test` command. We've previously mentioned that it is Go's test runner and that we will use it to execute tests. We'll look at how to use it in more detail in this section.

Inside the `_test.go` files, the test runner will treat three kinds of functions especially:

- **Test functions** that have a name beginning with `Test`. We have covered test functions at length in this section.
- **Benchmark functions** have a name beginning with `Benchmark`. We will cover these in the *Difference between a test and a benchmark* section of this chapter.
- **Example functions** that have a name beginning with `Example`. These are outside the scope of our discussion in this book.

The test runner will look for files that end in `_test.go`, build them into their own packages, and then link them in the test binary.

The output of the `go test` command will print out all the test failures of the executed tests to the standard output. You can add a `-v` flag, which is short for verbose, to get it to print the name and execution time of all tests, including passed tests.

Tests are executed and output in lexicographic order. Here is the output from our engine_test.go, which contains tests for the operations of the calculator, implemented in engine.go:

```
$ go test ./...
=== RUN    TestAdd
    engine_test.go:7: Add(2,3) incorrect, got: 2, want: 5.
--- FAIL: TestAdd (0.00s)
FAIL
exit status 1
FAIL    github.com/PacktPublishing/Test-Driven-Development-
in-Go/chapter02/calculator    0.278s
```

Test failures are marked in the output with the FAIL keyword and any error messages are printed to the standard output. In our example, we have a failing test: TestAdd. As we saw in the previous section, we can print informative error messages and fail tests using a variety of methods from the testing.T type, which we have access to in all tests from the parameter of the test signature.

At the end of the test output, we can see the outcome of the entire test run, as well as the time it took to run. We can also see the running time for each test.

The test runner supports two running modes:

- When the command has no package specifications, it will build and run all tests in the current directory. This is known as **local directory mode**. This is how we ran the preceding command using go test -v.

- When the command has package specifications, it will build and run all tests matching the specific package arguments. This is known as **package list mode**. Developers usually run their tests in this mode for large projects as it can be cumbersome to change between directories and run the tests in each of them using local directory mode.

We can easily specify what tests to run by providing these properties:

- **A specific package name**: For example, `go test engine_test` will run the tests from the `engine_test` package from anywhere in the project directory.

- **The expression as the package identifier**: For example, `go test ./...` will run all the tests in the project, regardless of where it's being run from.

- **A subdirectory path**: For example, `go test ./chapter02` will run all the tests in the `chapter02` subdirectory of the current path, but will not traverse to further nested directories.

- **A regular expression, together with the** `-run` **flag**: For example, `go test -run "^engine"` will run all packages that begin with the word `engine`. A subdirectory path can also be provided alongside the test name.

- **A test name, together with the** `-run` **flag**: For example, `go test -run TestAdd` will only the test specified. A subdirectory path can also be provided alongside the test name.

The Go test runner can cache successful test results to avoid wasting resources by rerunning tests on code that has not changed. Being able to cache successful test results is disabled by default when running in local directory mode, but enabled in package list mode.

As you can see, the output of the `go test -v ./...` command, which triggers package list mode, will mark the cached results with `(cached)` on their corresponding output line:

```
$ go test -v ./... === RUN    TestAdd
    engine_test.go:7: Add(2,3) incorrect, got: 2, want: 5.
--- FAIL: TestAdd (0.00s)
FAIL
FAIL    github.com/PacktPublishing/Test-Driven-Development-
in-Go/chapter02/calculator    0.112s
=== RUN    TestParser
--- PASS: TestParser (0.00s)
PASS
ok      github.com/PacktPublishing/Test-Driven-Development-
in-Go/chapter02/input          (cached)
FAIL
```

Note that only successful test runs can be cached. Test failures will be run every time until they pass, at which point they can be cached.

Writing tests

So far, we have examined the structure of packages, where test files fit into their structure, as well as become acquainted with Go's testing package and test signatures. Now, let's take everything we have learned and begin applying it.

With the knowledge we have of how to test code works in Go, we can expand on the **red, green, refactor** approach with more specific steps. *Figure 2.6* shows the expanded flow of TDD in Go for new functionality:

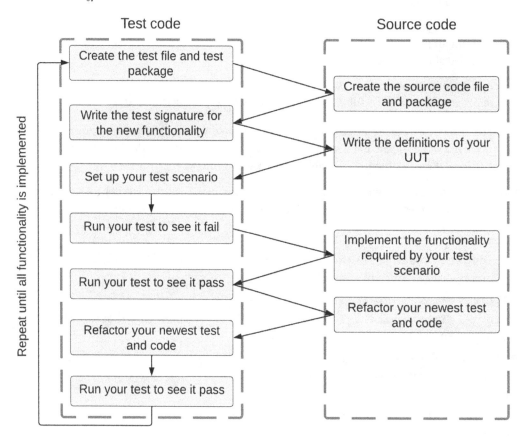

Figure 2.6 – Expanded TDD flow in Go

We can describe the test writing process with the following steps:

1. **Create the test file and test package**: You create a new directory to place your files and packages in. Then, you place the `_test.go` file corresponding to the test that you will be implementing in this new directory. This will ensure that you have a place to begin writing test code alongside implementation code. As previously mentioned, while Go does not enforce the usage of the external test package, you should use one whenever possible.

2. **Create the source code file and package**: In the same directory, create the source code file and declare the package name at the top to ensure that empty `.go` files are immediately compiled. At this point, you will begin reflecting on the structure of your code, as we did with the simple terminal calculator at the beginning of this chapter.

3. **Write the test signature for the new functionality**: While in the test file, you can create the new test, named after the UUT you will be testing. The test signature will also require you to import the testing package, getting you ready to write your test code.

4. **Write the definitions of your UUT**: Inside your previously empty source file, write the definitions for the custom types, methods, and functions that you intend to test. This will allow you to settle on the signatures or API of your UUT, then structure your test accordingly. As we did in the simple terminal calculator, return empty or zero values according to the signature of your methods to ensure your code compiles.

5. **Set up your test scenario**: Starting with the simplest test cases first, write your test using the AAA approach, invoking the previously defined signatures of your UUT. This is the reason we created the signatures and returned dummy values for the code to compile.

6. **Run your test to see it fail**: At this point, your code should be fully compiling so that you can run your tests to see the new code fail. To speed up the feedback further, you can run the `go test` command in the directory of your new package.

7. **Implement the functionality required by your test scenario**: Going back to your UUT, write just enough code to satisfy your newest test scenario. This will require changing some of your dummy code.

8. **Run your test to see it pass**: At this point, your code should be fully compiling so that you can execute your tests using the `go test` command. Your new test should now pass.

9. **Refactor your newest test and code**: Look for any improvements you can make to your source and test code. Improvements should be frequent and small, so make sure to take the time to review your code.

10. **Run your test to see it pass**: Your new test should continue to pass after this refactoring.

11. Repeat all of these steps as many times as required until all the functionality is implemented. You will define your UUT signatures and tests as required, starting from the simplest functionality and working forward.

As expected, the TDD process requires frequent changes between the source and test code. Test runs should always fail first, then pass as we implement and refactor the source code.

Use case – implementing the calculator engine

Let's use our established working practice to write and implement the functionality of the calculator that we will define in engine.go. This will also give us some hands-on experience with the testing package.

Step 1 – creating the test file and test package

As we saw in *Figure 2.2*, we will create a directory named calculator where we will place the corresponding calculator engine files.

We will create the engine_test.go test file and declare the external calculator_test package:

```
package calculator_test
```

At this point, the test file only contains a single line, and there are no compile errors.

Step 2 – creating the source code file and package

In the same directory, we must create the engine.go file and declare the calculator package, which matches the already declared external test package:

```
package calculator
```

At this point, the source code file also contains a single line and there are no compile errors.

Step 3 – writing the test signature for the new functionality

We will begin by testing and implementing the calculator addition functionality, as that is the simplest function. In the engine_test.go file, add a new test corresponding to the Go test signature and import the testing package:

```
package calculator_test

import "testing"

func TestAdd(t *testing.T) {
}
```

As its name and package indicate, we will be testing the Add function or method of the calculator package. With just these few lines of code, we already have a decent indication of what this test will cover. This is a very powerful mechanism.

Step 4 – writing the definitions of your UUT

Here, we will add stubbed definitions of the UUT to the source file so that we can invoke them in our newly written code. While this is a small departure from writing no code that does not have corresponding test code, it will make it easier for us to reference the code in test code from any code editor. In the engine.go file, we must add stubs for the Engine custom type and the Add method:

```
package calculator

type Engine struct {}

func(e *Engine) Add(x, y float64) float64{
  return 0
}
```

We return a dummy value of 0 to ensure that the code continues to compile.

Step 5 – setting up your test scenario

Going back to the test code, we will add a simple test scenario to the TestAdd function, which is currently empty. In the engine_test.go file, we will add testing code that was written using the AAA pattern, as indicated by the comments corresponding with each step:

```
package calculator_test

import (
  "testing"

  "github.com/PacktPublishing/Test-Driven-Development-in-Go/
chapter02/calculator"
)

func TestAdd(t *testing.T) {
  // Arrange
  e := calculator.Engine{}

  // Act
```

```
got := e.Add(2.5,3.5)


//Assert
if got != 6.0 {
  t.Errorf("Add(%.2f,%.2f) incorrect, got: %.2f, want:
    %.2f", 2.5, 3.5, got, 6.0)
  }
}
```

In the Arrange step, `TestAdd` creates an instance of `calculator.Engine`, which requires the calculator package to be imported inside the test file.

In the Act step, we invoke the `Add` method on the created `Engine` instance and pass it the two inputs that we will be using in this step.

Finally, in the Assert step, we compare the actual and expected values in an `if` statement and call for the test failure using `t.Errorf` if they do not match.

Step 6 – running your test to see if it fails

Starting with a failing test is very important to the TDD philosophy since it ensures that our test is actually being run and does not pass falsely. We can run our test:

```
$ go test -run TestAdd ./chapter02/calculator -v
--- FAIL: TestAdd (0.00s)
    engine_test.go:20: Add(2.50,3.50) incorrect, got: 0.00,
want: 6.00
FAIL
exit status 1
FAIL    github.com/PacktPublishing/Test-Driven-Development-
in-Go/chapter02/calculator    0.198s
```

The test fails and our informative error message is printed out to the terminal. The test is expected to fail since our method is currently only returning the dummy value. With that, we have completed the **red** phase of the red, green, refactor approach.

Step 7 – implementing the functionality required by your test scenario

With the failing `TestAdd` in place, it's time to implement the functionality required to make it pass. In the `engine.go` file, we must change the `Add` method to remove the dummy value return:

```go
func(e *Engine) Add(x, y float64) float64{
   return x + y
}
```

The `Add` method will now use the input parameters and return their addition result.

Step 8 – running your test to see if it passes

Exactly like in *Step 6 – running your test to see if it fails*, we will run the test:

```
$ go test -run TestAdd ./chapter02/calculator -v
=== RUN    TestAdd
--- PASS: TestAdd (0.00s)
ok         github.com/PacktPublishing/Test-Driven-Development-
in-Go/chapter02/calculator    0.109s
```

The test is now passing and we have verified that our code satisfies the test requirements. With that, we have completed the **green** phase of the red, green, refactor approach.

Step 9 – refactoring your newest test and code

This step is not guaranteed to take place. We can improve our test code by extracting variables that allow us to clean up the test code, removing the need for us to repeat hardcoded values:

```go
func TestAdd(t *testing.T) {
   // Arrange
   e := calculator.Engine{}
   x, y := 2.5,3.5
   want := 6.0

   // Act
   got := e.Add(x,y)

   //Assert
   if got != want {
      t.Errorf("Add(%.2f,%.2f) incorrect, got: %.2f, want: %.2f",
            x, y, got, want)
```

```
   }
}
```

In the Arrange section, we now declare three variables for our inputs and expected output. We make use of these variables throughout the test, passing them to the UUT as well as to the formatted error message.

Step 10 – re-running your test to see if it passes

Exactly like in *Step 8 – running your test to see if it passes*, we must run the test:

```
$ go test -run TestAdd ./chapter02/calculator -v
=== RUN    TestAdd
--- PASS: TestAdd (0.00s)
ok      github.com/PacktPublishing/Test-Driven-Development-
in-Go/chapter02/calculator    0.196s
```

The test is now passing and we have verified that our refactoring has not broken any implemented functionality. With that, we have completed the **refactor** phase of the red, green, refactor approach.

These steps can now be repeated for all the other operations of the simple terminal calculator. You can go ahead and implement them, which will allow you to practice using TDD in Go. Next, we'll explore how to streamline our test writing processing using test setup and teardown.

Test setup and teardown

We've written our first example of a test and source code by leveraging an external test package and the `testing.T` type. This has worked very well for our simple example, but, as we begin to ramp up and write more tests, it can be cumbersome to continue repeating the same test setup and cleanup. In this section, we will explore what functionality the `testing` package offers to streamline this process for us.

The TestMain approach

Go 1.4 introduced a new special test called `TestMain`. This is a feature that is often underutilized, but it gives us great flexibility when it comes to setup and teardown code. The signature of this test is as follows:

```
func TestMain(m *testing.M) {
   // implementation
}
```

Unlike other tests, the name of this test is fixed and it takes in the *testing.M type as its only parameter, as opposed to *testing.T as other tests do. Once you override it, the code in this method will give you more control over how your tests run. The TestMain method will be run before any of the other tests in this package.

> **One TestMain function per package**
>
> As names need to be unique inside a package, you will only be able to define one TestMain function per package. You should be mindful that this method will control how all the tests inside the given package run, not just those in the given file.

The testing.M type is much smaller than the testing.T type and exposes a method called Run(), which allows us to run the tests in the given package and returns an exit code.

The usage of this function is simple:

```go
func TestMain(m *testing.M) {
    // setup statements
    setup()

    // run the tests
    e := m.Run()

    // cleanup statements
    teardown()

    // report the exit code
    os.Exit(e)
}
```

The preceding code sample outlines a simple recipe:

1. **Declare the special** TestMain **signature**: Write the correct name and signature of the test in your test file. In general, you should place this definition as high up at the beginning of the file as possible.

2. **Write your setup code**: Inside the body of the main test, you should write your setup code. I recommend writing a separate setup() function and calling that instead of writing the code directly into your test function. This will help with the readability of your test file. All of these statements will run before your tests are executed.

3. **Invoke the** Run () **function**: After you have written your setup code, you will have to invoke m.Run () and save the exit value returned from this function inside a variable, named e in our code sample. It is at this point that the tests will run and the exit value will report whether your tests have failed.

4. **Write your teardown/cleanup code**: In the same way as the setup, you can write your teardown or cleanup code after the invocation of the Run () method. I also recommend creating a separate function named teardown, as opposed to writing the code directly into your TestMain code block. All of these statements will run after your tests are executed.

5. **Report the exit value**: This step is very important as it allows us to indicate test failures to the test runner. You should pass the exit code returned from the test run and pass it to the os.Exit function. If you forget to add this part of the main function, you might get false positives reported to your test runner.

We implement the same recipe in our calculator example, defining the TestMain function alongside TestAdd in the engine_test.go file:

```go
func TestMain(m *testing.M) {
    // setup statements
    setup()

    // run the tests
    e := m.Run()

    // cleanup statements
    teardown()

    // report the exit code
    os.Exit(e)
}

func setup() {
    log.Println("Setting up.")
}

func teardown() {
    log.Println("Tearing down.")
}
```

The setup() and teardown() functions simply print two lines to the Terminal. Running the test shows us the following output:

```
$ go test -run TestAdd ./chapter02/calculator -v
2022/08/14 11:02:51 Setting up.
=== RUN    TestAdd
--- PASS: TestAdd (0.00s)
PASS
2022/08/14 11:02:51 Tearing down.
ok      github.com/PacktPublishing/Test-Driven-Development-
in-Go/chapter02/calculator    0.345s
```

As we can see from the test output, the setup and teardown log lines are printed around the test run output, before and after.

init functions

The second option you have available to ensure that the test setup runs correctly is to use init functions. It is often the case with unit tests that no teardown logic is required, only setup logic. In these cases, where you simply want to ensure that some logic is run before the tests, you might want to opt for a less cumbersome approach than TestMain.

Unlike the TestMain approach, the init functions are not specifically restricted to test code. The signature of the init function looks like this:

```
func init() {
    // implementation
}
```

The name of the init function is fixed and it takes no parameters. This function will be called before any main function, regardless of whether that main function is in the source code or the special test runner main function.

> **Multiple init functions per package**
>
> Unlike other names, multiple init functions are allowed per package. However, you should be mindful that they will all be called before the main runner. When multiple init functions are defined in the same file, they are run in definition order. On the other hand, when they are defined in multiple files, they are run in the lexicographic order of their filenames.

We will define an init function, alongside `TestMain` and `TestAdd`, in the `engine_test.go` file:

```
func init() {
  log.Println("Init setup.")
}
```

The `init()` function simply prints another line to the Terminal. Running the test shows us the following output:

```
$ go test -run TestAdd ./chapter02/calculator -v
2022/08/14 11:57:38 Init setup.
2022/08/14 11:57:38 Setting up.
=== RUN    TestAdd
--- PASS: TestAdd (0.00s)
PASS
2022/08/14 11:57:38 Tearing down.
ok       github.com/PacktPublishing/Test-Driven-Development-
in-Go/chapter02/calculator    0.252s
```

As we can see from the test run output, the init setup is run before the `TestMain` setup. The logline defined in the init function is printed before any other code is executed.

Deferred functions

We can make use of **deferred functions** for teardown logic that does not leak outside the scope of the current test. Just like the init functions, this is a construct that does not only exist in test code.

In Go, deferred functions are declared using the `defer` statement. Once this is applied to a function call, the function will only be executed once the surrounding function call has been completed, either successfully or using a panic. For example, we can defer the teardown function like so:

```
defer teardown()
```

> **Deferred function definitions**
>
> We can apply the `defer` statement to named functions or **anonymous functions** defined inline. It is Go convention to define your deferred functions at the top of the enclosing function. This will ensure that the function will be deferred before any errors can occur and stop the deferral.

The approaches we've seen so far are made up of Go's language constructs, but they can be cumbersome to keep defining and have the disadvantage of creating package-level changes. Deferred functions give us fine-grained control, making changes only to the test where they are invoked. However, the disadvantages are that we need to remember to add them to each test and that we can only use this approach for teardown, not setup logic. You should weigh the advantages and disadvantages of each mechanism as you begin to write more tests.

Let's modify the TestAdd function in the engine_test.go file to add a deferred function, leaving the TestMain and init functions already defined in this test file unchanged:

```
func TestAdd(t *testing.T) {
  defer func ()  {
  log.Println("Deferred tearing down.")
}()

  // Arrange
  e := calculator.Engine{}
  x, y := 2.5, 3.5
  want := 6.0

  // Act
  got := e.Add(x, y)

  //Assert
  if got != want {
    t.Errorf("Add(%.2f,%.2f) incorrect, got: %.2f, want:
      %.2f", x, y, got, want)
  }
}
```

The deferred function simply prints out another log line to the Terminal. Running the test shows us the following output:

```
$ go test -run TestAdd ./chapter02/calculator -v
2022/08/14 12:25:49 Init setup.
2022/08/14 12:25:49 Setting up.
=== RUN   TestAdd
2022/08/14 12:25:49 Deferred tearing down.
--- PASS: TestAdd (0.00s)
PASS
2022/08/14 12:25:49 Tearing down.
```

```
ok        github.com/PacktPublishing/Test-Driven-Development-
in-Go/chapter02/calculator    0.215s
```

As we can see from the test run output, the deferred teardown call is executed before the `TestMain` function's teardown step. This is expected due to the invocation order of deferred functions.

Figure 2.7 depicts a summary of the order in which all the setup and teardown mechanisms we've looked at will execute:

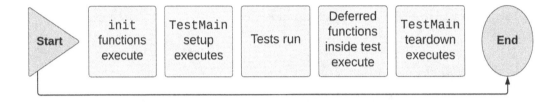

Figure 2.7 – Summary of the order of setup and teardown mechanisms

The order verifies what we've seen with our terminal output:

1. The tests kick off with the `go test` command, as we've become used to running them so far.
2. The `init` functions execute before the temporary main program of the tests.
3. Once the tests are ready to execute, the `TestMain` function starts and its setup functions execute.
4. The tests are then run by invoking `m.Run()` from `TestMain`.
5. Once all the tests have been run, the deferred functions defined inside the scope of the tests are executed.
6. Once the tests and their functions exit, the `TestMain` function's teardown function is executed.
7. Finally, the tests end with the exit value returned from the call to `m.Run()`.

As we begin to consider writing tests on a larger scale, we will also need a way to separate tests according to a smaller test scope and different scenarios. In the next section, we'll see how to achieve that using **subtests**.

Operating with subtests

In TDD, the test scope should be small, and the outcome should be easy to understand. We explored TDD best practices in *Chapter 1, Getting to Grips with Test-Driven Development*. To achieve these best practices, we need separation between test scenarios.

Let's consider the example of the `TestAdd` function that we have worked with so far. It currently tests adding two positive numbers, but we will extend it to cover negative inputs. With the knowledge we have so far, we have two options:

1. **Extend the scope of `TestAdd` to verify the new scenario**: This approach will make the Assert step longer, but it has the advantage of reusing the previous steps.

2. **Create a new test to verify the new scenario**: This approach will keep the scope of `TestAdd` as is, but it has the disadvantage of us having to redefine and re-execute the Arrange and Act steps of the existing test.

If we choose to go with the second option, we will have to name the new test something different. We will name it `TestAdd_Negative` to signify that we will be testing negative inputs in this test. However, this doesn't fall in line with the naming of the existing `TestAdd` function, so we will have to rename the existing test to `TestAdd_Positive`. As expected, running the tests will output the result on different lines:

```
$ go test -run "^TestAdd" ./chapter02/calculator -v
=== RUN    TestAdd_Positive
--- PASS: TestAdd_Positive (0.00s)
=== RUN    TestAdd_Negative
--- PASS: TestAdd_Negative (0.00s)
PASS
ok      github.com/PacktPublishing/Test-Driven-Development-
in-Go/chapter02/calculator    0.266s
```

We want to have a small self-contained test, but it would be cumbersome to continue to define new tests, potentially having to change the name of existing tests for each new edge case or scenario we test. Go provides us with a more elegant solution to this common problem, which we will cover in the next section.

Implementing subtests

The `testing.T` type provides the `Run(name string, f func(t *testing.T)) bool` method, which takes in two parameters:

- A name parameter of the `string` type
- A function that takes in a single parameter of the `*testing.T` type

Once passed to the Run method, the test runner will run the function as a subtest of the current tests, allowing us to create a test hierarchy, each with its own separation. Since the enclosing test and the subtests share the same instance of testing.T, a subtest failure will cause the enclosing test to fail as well. This behavior gives us the ability to create multi-layered test hierarchies according to our needs. Taking the example of adding positive and negative inputs as test scenarios, we can refactor TestAdd to take advantage of the power of subtests:

```go
func TestAdd(t *testing.T) {
  // Arrange
  e := calculator.Engine{}

  actAssert := func(x, y, want float64) {
  // Act
  got := e.Add(x, y)

  //Assert
  if got != want {
    t.Errorf("Add(%.2f,%.2f) incorrect, got: %.2f, want:
      %.2f", x, y, got, want)
  }
}

  t.Run("positive input", func(t *testing.T) {
    x, y := 2.5, 3.5
    want := 6.0
    actAssert(x, y, want)
  })

  t.Run("negative input", func(t *testing.T) {
    x, y := -2.5, -3.5
    want := -6.0
    actAssert(x, y, want)
  })
}
```

We create an `actAssert` function that takes in the inputs and expected output as parameters. This function will perform the Act and Assert steps without having to repeat them. Then, we create two subtests using the `t.Run` method we've seen before. The name of each subtest indicates what scenario it will cover. Running the tests will produce the following result:

```
$ go test -run "^TestAdd" ./chapter02/calculator -v
=== RUN    TestAdd
=== RUN    TestAdd/positive_input
=== RUN    TestAdd/negative_input
--- PASS: TestAdd (0.00s)
    --- PASS: TestAdd/positive_input (0.00s)
    --- PASS: TestAdd/negative_input (0.00s)
PASS
ok      github.com/PacktPublishing/Test-Driven-Development-
in-Go/chapter02/calculator    0.195s
```

As we can see from the output, the subtests are nested under the enclosing test. By leveraging subtests, we now have a convenient way to create tests that share the Arrange step, but can also be easily extended with more scenarios without the need to rename tests.

We will discuss the related technique of **table-driven testing**, which leverages the power of subtests, in *Chapter 4, Building Efficient Test Suites*.

Code coverage

Now that we know how to write tests that cover different scenarios and how to run them, we can have a look at what our **code coverage** is. As we remember from *Chapter 1, Getting to Grips with Test-Driven Development*, this important metric measures what percentage of your code is exercised by tests.

The `go test` command has a `-cover` flag, which computes the code coverage profile of the given package. It also offers the possibility of saving the profile to a file by passing a file path to the `-coverprofile` flag. We will then see how to view these saved coverage profiles.

Let's run it for our calculator:

```
$ go test -run "^TestAdd" ./chapter02/calculator -cover -v
=== RUN    TestAdd
=== RUN    TestAdd/positive_input
=== RUN    TestAdd/negative_input
--- PASS: TestAdd (0.00s)
    --- PASS: TestAdd/positive_input (0.00s)
    --- PASS: TestAdd/negative_input (0.00s)
```

```
PASS
coverage: 100.0% of statements
ok      github.com/PacktPublishing/Test-Driven-Development-
in-Go/chapter02/calculator    0.113s
```

This command prints out the coverage percentage after running all the tests. We are currently measuring coverage of 100% since our Add function is very simple.

Now, let's save the code coverage profile to a file using go test ./chapter02/calculator -coverprofile=calcCover.out. This will create the calcCover.out file in the current directory. We can view this file visually using another tool in the Go toolchain. Running go tool cover -html=calcCover.out will open a new window in your browser to display the coverage profile visually.

Figure 2.8 shows the visual representation of our cover profile, which shows that the Add method is covered by tests:

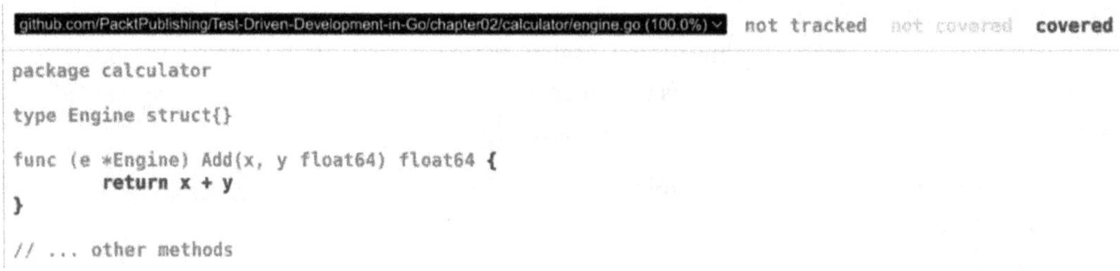

github.com/PacktPublishing/Test-Driven-Development-in-Go/chapter02/calculator/engine.go (100.0%) ⌄   not tracked   not covered   **covered**

```
package calculator

type Engine struct{}

func (e *Engine) Add(x, y float64) float64 {
        return x + y
}

// ... other methods
```

Figure 2.8 – The visual representation of the saved profile

That covers all the essentials we need to know to begin writing Go tests with TDD. The last thing we need to tackle is how to write and use benchmarks.

The difference between a test and a benchmark

The last concept we will be looking at in this chapter is **benchmarks**. The testing package gives us access to benchmarks using the testing.B type. They have a signature very similar to tests:

```
func BenchmarkName(b *testing.B) {
    // implementation
}
```

The signature highlights the following requirements for Go benchmarks:

- Benchmarks are exported functions whose name begins with Benchmark.

- Benchmark names can have an additional suffix that specifies what the test is covering. The suffix must also begin with a capital letter, as we can see with Name.

- Benchmarks must take in a single parameter of the *testing.B type. As we've explained so far, this will be how the test interacts with the test runner. You can name the testing parameter however you want, but Go developers typically use b to denote it.

- Benchmarks must not have a return type.

> **Benchmarks are an important Go profiling tool**
>
> Tests verify the functionality of your programs, while benchmarks verify the performance of your code. You should use both in your testing strategy.

Benchmarks can also be run with the go test command, but we have to specify to the runner that we are interested in benchmarks with the –bench flag. We must supply a regular expression that matches the packages that we want to run. We can run all benchmarks by matching all packages in the current directory using this command:

```
go test -bench. .
```

The testing.B type also has access to logging errors and signaling test failures, just as we saw in the introduction to the testing.T type: b.Error, b.Errorf, b.Fatal, and b.Fatalf. Just like tests, benchmarks live in test files, which must have the _test.go suffix to be detected by the Go test runner.

Let's write a benchmark for our Add function in the engine_test.go file:

```go
func BenchmarkAdd(b *testing.B) {
  e := calculator.Engine{}

  // run the Add function b.N times
  for i := 0; i < b.N; i++ {
    e.Add(2, 3)
  }
}
```

The BenchmarkAdd example runs the Add function with the parameters in a loop for b.N times. Go's test runner controls the value of N and will increase it until it is satisfied that the numbers it has measured are stable.

As with all performance tests, you should be wary of running benchmarks on your local machine. You might measure some large variations in measurement according to what your computer is processing.

Now, we run our benchmark to see the following output:

```
$ go test -bench. ./ chapter02/calculator -v
pkg: github.com/PacktPublishing/Test-Driven-Development-in-Go/
chapter02/calculator
cpu: IntelI Core(TM) i5-8279U CPU @ 2.40GHz
BenchmarkAdd-8                 1000000000                 0.2684 ns/op
PASS
ok        github.com/PacktPublishing/Test-Driven-Development-
in-Go/chapter02/calculator    0.408s
```

The output of the benchmark run highlights the following:

- The name of the benchmark: BenchmarkAdd

- The number of CPU cores used to run the benchmark, added as a suffix to the test name: 8

- How many times the benchmark was executed: 1000000000

- The average amount of time that an individual test iteration took, measured in nanoseconds: 0.2684 ns/op

Our function is very simple, which is why it has a very low running time. We will explore more complex examples of benchmarks in *Chapter 8, Testing Microservice Architectures*.

Summary

In this chapter, we covered all the unit testing essentials that we will need to get started with TDD in Go. We began by introducing Go modules and packages, as well as where test files are placed and how they declare their additional test packages. You learned about the most important methods in Go's testing package and started writing some simple unit tests with it. Then, we explored ways to reduce code duplication by making use of setup and teardown functions, as well as how to better organize tests using subtests. Finally, we learned how to write and run benchmarks, which are an important part of any testing strategy.

In *Chapter 3, Mocking and Assertion Frameworks*, we will write more complicated tests, which require dependencies. We will explore some popular frameworks and begin to use them to write tests that are closer to real-world examples.

Questions

Answer the following questions to test your knowledge of this chapter:

1. In Go, what is the difference between a module and a package?
2. What is the additional test package? What are some of the advantages of using it?
3. What are the requirements for the test signature?
4. What are subtests and how do you create them?
5. What is a benchmark? How do you write one?

Further reading

To learn more about the topics that were covered in this chapter, take a look at the following resources:

*   The official documentation for the testing package is available at `https://pkg.go.dev/testing`
*   *The Art of Unit Testing*, by Roy Osherove, published by Manning Publications
*   *Profiling Go Programs* is available on the Go blog at `https://go.dev/blog/pprof`

3

Mocking and Assertion Frameworks

In the previous chapter, we explored the fundamentals of writing tests in Go. We explored the importance of packages, the organization of test files alongside source code, and how to use Go's `testing` package to write tests and benchmarks.

We demonstrated the concepts and fundamentals of writing tests in Go with code samples from the `Calculator` use case. The simple examples we have looked at so far have not included any external dependencies, which can make test setup and verification much more complex.

In this chapter, we will begin to look at how we can isolate the unit under test from its dependencies, keeping testing and assertions as simple and fast as possible. The easiest way to achieve this in Go is by leveraging the power of interfaces.

We will expand upon the `Calculator` example by introducing dependencies to our main components. Then, we will learn how to generate mocks for these dependencies of the **unit under test** (**UUT**), enabling us to control their behavior.

Then, we will explore some external, open source assertion libraries that are often used by Go engineers. Up until now, we have written some simple assertions ourselves. This can be repetitive and limiting. The `testify` and `ginkgo` assertion libraries are two popular choices that can be used to supplement Go's `testing` package.

Finally, we will turn away from exploring test code and mechanisms and look at design techniques for writing testable code by reviewing the **SOLID** principles of object-oriented design. We will learn what they are and how they can be applied when writing code.

In this chapter, we will cover the following topics:

- Using interfaces to wrap dependencies

- How to generate and use mocks to test code in isolation

- The usage of popular assertion frameworks

- What the **SOLID** design principles are

- Best practices for writing testable code

Technical requirements

You will need to have **Go version 1.19** or later installed to run the code samples in this chapter. The installation process is described on the official Go documentation at `https://go.dev/doc/install`.

The code examples included in this book are publicly available at `https://github.com/PacktPublishing/Test-Driven-Development-in-Go/chapter03`.

Interfaces as dependencies

As always, implementing and exploring unit testing techniques begins with exploring code writing techniques. This is a theme we will see regularly throughout this book. We cannot study testing in isolation. It requires insight into the code design and its intended purpose.

In this section, we will look at the concept of **software dependencies** and how to manage them. *Figure 3.1* depicts the three main types of dependencies:

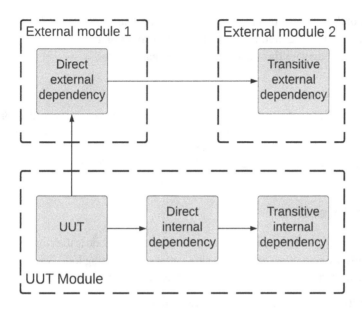

Figure 3.1 – Types of dependencies from the point of view of the UUT

From the viewpoint of the UUT, the four main types of dependencies are as follows:

- **Direct internal dependencies**: These contain internal functionality that your **UUT** imports. These dependencies could be defined in the same package or module as **UUT**, but are required to deliver its functionality.

- **Transitive internal dependencies**: These contain internal functionality that the **Direct internal dependency** parts of your **UUT** import. These dependencies could also be defined in the same package or module.

- **Direct external dependencies**: These contain third-party functionality that your **UUT** imports. These could be libraries or service APIs that you might rely on, but which are not contained in your current module.

- **Transitive external dependencies**: These contain external functionality that your **Direct external dependencies** rely on, but which are in a separate module. Due to the way that Go builds the source code and required libraries into runnable executables, these transitive dependencies will also be contained alongside your code during application release.

The dependencies of the UUT are required for the UUT to be able to correctly deliver its functionality. Consequently, they are also required to completely test its functionality. We will explore techniques for handling code dependencies throughout this section and chapter.

> **Don't reinvent the wheel**
>
> Writing code that relies on dependencies is a normal, and recommended, practice for software design. It allows us to reuse behavior and implementation in multiple places. This, coupled with Go's powerful module and package system, makes it easy and fast to write complex code. We explored Go's modules and packages in *Chapter 2, Unit Testing Essentials*.

Dependency injection

One popular and common technique for handling dependencies in code is the concept of **dependency injection (DI)**. This is a simple yet effective design pattern for creating loosely coupled code, which allows us to implement code without the concerns of its dependencies.

DI is a style of writing code in which the UUT or function receives other types or functions that it depends on during initialization. Fundamentally, DI is nothing more than passing the correct parameters to a function, which then uses these to create the UUT.

This technique is one of the principles of SOLID design, namely the letter *D*, which stands for the principle of **dependency inversion**. We will explore all the principles later in this chapter, in the *Writing testable code* section.

> **Why is it called injection?**
>
> The term *injection* simply signifies that the dependencies are not created by the UUT that requires them but passed to it from further up the stack. They can be injected either by constructor/function injection or by the use of frameworks.

Figure 3.2 describes the main steps of what DI typically entails:

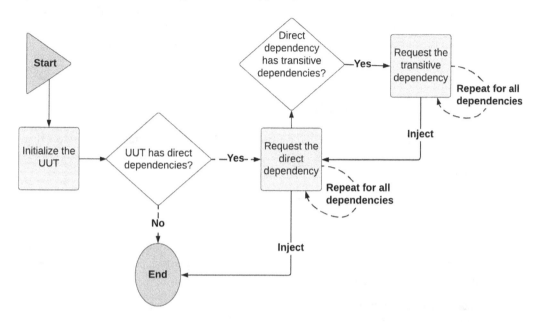

Figure 3.2 – The main steps of dependency injection

We can see the following sequence of calls:

1. At the start, we begin by attempting to **Initialize the UUT**. In Go, the UUT is usually a `struct`. We already know that Go structs do not provide constructors, so the initialization process will involve inspecting the dependencies that the UUT requires. We will see techniques for making the need for dependencies explicit in the next section.

2. If the UUT has any direct dependencies, then we will **Request the direct dependency**. This direct dependency will also typically be a `struct`, either from the same module or another external module.

3. When initializing the direct dependency, we might discover transitive dependencies that the direct dependencies need during initialization. We will then **Request the transitive dependency**.

4. The dependency request process is repeated for all direct and transitive dependencies.

5. Once each dependency has been successfully created, it is injected into the creation of the previous dependency or the UUT, if it is a direct dependency.

The dependency graph

Due to the way that dependencies need to be created and then, in turn, injected, this process is known as *constructing the dependency graph*. This is a directed, acyclic graph. This graph allows the compiler to start at the root and then traverse it while building all the custom types it requires while running the main program.

Implementing dependency injection

While introducing DI, we briefly mentioned that structs do not have constructors and therefore this process might require the investigation of the properties and fields of a `struct`. Let us have a look at how we can implement DI in Go.

Fundamentally, there are two ways we can go about injecting dependencies:

- **Constructor injection**: This consists of passing all the required dependencies to a special constructor function, which will then return an instance of the UUT `struct`. This is an intuitive way to construct instances, but it does require that all dependencies be created before the invocation of the function.

- **Property/method injection**: This consists of creating the UUT `struct` and then setting the fields of the dependencies as you require them. This can either be done by directly setting them as fields on the UUT instance, or by invoking setter methods that set them on the fields. The dependencies are not immutable, so they do not require the UUT instance to be recreated as they are set. This way of creating the UUT and its dependencies does not require that all dependencies be created before initializing and beginning to use the UUT, but it also does not guarantee that all the dependencies will be set by a certain time, nor does it guarantee they won't be changed later. This could require more application code for `nil` value checks, as well as other subtle bugs if dependencies change.

Then, each method can be used in two ways:

- **Manual invocation**: This means that we call and create the UUT struct and its dependencies manually. In this process, you have full control over the creation and invocation of dependencies, but they can become difficult to manage for larger codebases.

- **Dependency injection frameworks**: This means that you import another dependency into your project that can automate this process using advanced techniques such as reflection or code generation, which then leverage the dependency graph to create the dependencies in the correct sequence. This method is much more sustainable for large codebases.

When it comes to DI frameworks, there are two popular open source choices that you can use in your code:

- `dig`: This uses reflection to build your dependency graph and successfully construct your instances. You can read about how to use it at `https://github.com/uber-go/dig`.

- `wire`: This uses reflection and code generation for DI. You can read about how to use it at `https://github.com/google/wire`.

> **Reducing complexity**
>
> Remember that one of the core principles of Go code and software design is simplicity. You should keep your code as simple as possible, avoiding the lengthy constructors that have been seen in other legacy languages.

When it comes to dependencies, they are usually represented using corresponding interface types. This is an approach unique to Go, regardless of how you choose to inject your dependencies. Let us look at their role in software design a little bit more closely.

Interfaces are named collections of zero or more methods. Here are some key highlights of their behavior:

- The y are the primary way we can achieve **polymorphism** is in Go.

- The compiler enforces them and implicitly casts a `struct` to its corresponding interface.

- To implement an interface, a `struct` needs to implement its defined methods.

- A `struct` can implement multiple interfaces, so long as it satisfies its method signatures.

- An interface with zero methods is the empty interface and its type is `interface {}`. This is useful in certain cases, but the interfaces you create will have one or more methods.

- The zero value of interfaces is nil. We will need to handle this in our code as we begin to use interfaces to wrap around our dependencies.

> **Interfaces define methods, not functions**
>
> Remember that interfaces define methods, not functions. As we've seen with the `Engine` definition, methods that correspond to the signature of the interface will need to be defined on the struct we want to use in place of this interface.

Let us look at a DI example; this can be found in the `chapter03/di/manual/calculator.go`. We can define a simple `Adder` interface with the following snippet:

```
type Adder interface {
   Add(x, y float64) float64
}
```

The `Adder` interface defines the `Add` method. Note that this method takes in two `float64` parameters and returns one `float64` return value. In our case, `Engine` will satisfy this interface as it implements this method:

```
func (e Engine) Add(x, y float64) float64 {
   return x + y
}
```

When we initialize `Engine`, we can return the `Engine` struct:

```
func NewEngine() *Engine {
   return &Engine{}
}
```

A simple `Calculator` then makes use of this `Engine` for its adder functionality and prints out the result:

```
type Calculator struct {
  Adder Adder
}
func NewCalculator(a Adder) *Calculator {
   return &Calculator{Adder: a}
}
func (c Calculator) PrintAdd(x, y float64) {
   fmt.Println("Result:", c.Adder.Add(x, y))
}
```

`Engine` is a dependency of `Calculator`, so it is a parameter of the `NewCalculator` function. `Adder` is then invoked inside the `PrintAdd` method, where its functionality is required. Therefore, the initialization process of `Calculator` requires an instance of `Engine` to be created to compile:

```
func main() {
   engine := NewEngine()
   calc := NewCalculator(engine)
```

```
    calc.PrintAdd(2.5, 6.3)
}
```

This example uses the *manual invocation* of DI. As the dependency graph grows in size and complexity, this initialization function will become increasingly cumbersome and require changes. This is where DI frameworks can help simplify our code.

Using the previously introduced `wire` framework, we can define an `InitCalc` function in the `/chapter03/di/wire/wire.go` file, which will take care of initializing `Calculator` with its Engine:

```
//go:build wireinject
package main
import "github.com/google/wire"
var Set = wire.NewSet(NewEngine, wire.Bind(new(Adder),
new(*Engine)), NewCalculator)
func InitCalc() *Calculator {
   wire.Build(Set)
   return nil
}
```

This `wire.Build` function takes in the `Set` that matches the `Adder` interface to the `Engine` struct. At the top of the file, we make use of a build tag to exclude this file from the final binary, and use the generated replacement file when we run our application.

Next, we must install the wire tool and run it in the correct directory:

```
$ go install github.com/google/wire/cmd/wire@latest
$ cd chapter03/di/wire && wire
wire: github.com/PacktPublishing/Test-Driven-Development-in-Go/
chapter03/di/wire: wrote /Users/adelinasimion/code/Test-Driven-
Development-in-Go/chapter03/di/wire/wire_gen.go
```

This command generates the `wire_gen.go` file, which contains the implementation of the `InitCalc` function:

```
func InitCalc() *Calculator {
   adder := NewEngine()
   calculator := NewCalculator(adder)
   return calculator
}
```

This function contains the dependency creation code that we had previously written by hand. As it is now maintained and generated by wire, changes will not have to be maintained manually and the main function is now simpler:

```
func main() {
  calc := InitCalc()
  calc.PrintAdd(2.5, 6.3)
}
```

Finally, we can build the application, generating the initialization function and binding it into the Go binary. Then, we can run the executable as usual:

```
$ go build./chapter03/di/wire
$ ./wire
Result: 8.8
```

DI frameworks simplify the code we write and maintain but do require adding new steps to the build process, as well as an extra cognitive load when first starting with them. We explored how the `wire` DI library works in this section, but we will be using manual injection going forward so that we have more control and can explore the code together better.

Use case – continued implementation of the calculator

In this section, we will make use of the techniques we have seen so far to expand upon the implementation of the calculator from *Chapter 2, Unit Testing Essentials*.

Setting aside the correct procedure of **test-driven development** (TDD), let us consider this sketched implementation of the `input.Parser` struct:

```
type Parser struct {
  engine    *calculator.Engine
  validator *Validator
}


// ProcessExpression parses an expression and sends it to
// the calculator
func (p *Parser) ProcessExpression(expr string) (*string,
error) {
  operation, err := p.getOperation(expr)
  if err != nil {
    return nil, format.Error(expr, err)
```

```
    }
    return p.engine.ProcessOperation(*operation)
}
```

As we know from *Chapter 2*, *Unit Testing Essentials*, where we first looked at this example, Parser depends on Validator and calculator.Engine. These two structs are the direct dependencies of Parser. Then, these dependencies are used to deliver the functionality of the ProcessExpression method.

Regardless of whether we use third-party DI frameworks or we create the corresponding structs manually, writing tests for this relatively simple code snippet involves:

- Initializing the Parser struct with all its direct and transitive dependencies. This could involve a lengthy setup, with external dependencies that might extend the scope of the test.

- Once these main building blocks have been created, we need to set up their pre-condition state. This could involve an even more complicated setup, which could have unintended consequences as well.

- When it comes to verification, we might need to assert the internal state of dependencies to ensure that they are behaving as expected. This reliance on the internal state of dependencies would then make the tests more brittle since changes to the dependencies would break the tests.

Now that we understand how to build code that requires direct dependencies, we will begin to explore mechanisms that can help us in testing such dependencies. We will leverage Go development tools to make testing and assertions easier.

> **Controlling test scope**
>
> When it comes to types that have many dependencies, the scope of the test setup and assertion can increase beyond the UUT. We need a mechanism that allows us to test the UUT in isolation, which also has the benefit of keeping the test scope small.

Exploring mocks

In this section, we will explore one of the mechanisms that allows us to test code that relies on dependencies. We will see how to use and generate mocks, allowing us to verify the UUT in isolation from the behavior of its dependencies.

Mocks are sometimes known as **test doubles** and are a simple but powerful concept. They satisfy the interfaces but are fake versions of the real implementations. We have full control over these fake implementations, giving us the freedom to control their behavior. However, if the real implementation changes and our mocks do not, then our tests will give us false confidence.

In Go, we have the following different mocking options:

- **Function substitution**: This means sending replacement fake functions to the UUT. This is easy to do in Go, which has native support for **higher-order functions**. We can override function variables and parameters, replacing the behavior of the UUT.

- **Interface substitution**: This means injecting fake versions of the interfaces that the UUT depends on. These are fake stubbed implementations that satisfy the interfaces of the real implementation. They can then be used to replace the full implementations, without the UUT even being aware of it.

> **Higher-order function refresher**
>
> A higher-order function is a function that either takes in another function as a parameter or returns a function. Go functions are just like any other type.

The use of function substitution is less prevalent than interface substitution and should be used sparingly since it can make code less readable.

Now, let us change our code to be able to take advantage of **interface substitution**. First, we will define the two interfaces that we will be calling:

```go
// OperationProcessor is the interface for processing
// mathematical expressions
type OperationProcessor interface {
  ProcessOperation(operation *calculator.Operation) (*string,
error)
}


// ValidationHelper is the interface for input validation
type ValidationHelper interface {
  CheckInput(operator string, operands []float64) error
}
```

The following points describe the preceding code:

- We begin by defining interfaces for the external functionality that we want to leverage in the UUT. In our case, the UUT is input.Parser and we will need two dependencies:

  - The OperationProcessor interface wraps around the ProcessOperation method. This functionality will be satisfied by calculator.Engine and will calculate the mathematical result of the parsed operator and operands.

  - The ValidationHelper interface wraps around the CheckInput method. This functionality will be satisfied by input.Validator and will ensure that the user-supplied input can be processed.

> **Exported dependency interfaces**
>
> Note that the interfaces wrapping around dependencies have been exported, as can be seen by their capitalized names. It is common practice for interfaces to be exported and their corresponding structs to stay inside the package scope. This allows us to have fine-grained control over what functionality is exposed outside of the current package.

Then, we wrap the dependencies of the input.Parser type with the newly defined interfaces:

```
// Parser is responsible for converting input to
// mathematical operations
type Parser struct {
    engine    OperationProcessor
    validator ValidationHelper
}
```

As we discussed in the previous section, *Interfaces as dependencies*, Go dependencies are usually represented as interfaces, instead of struct types. This allows us to inject any type that satisfies the given interface, as opposed to only the concrete struct. This is a very powerful mechanism.

Another big advantage of using interfaces to represent dependencies is that they allow us to break the dependencies between packages, and write **loosely coupled code**.

Figure 3.3 depicts how we can break hard dependencies:

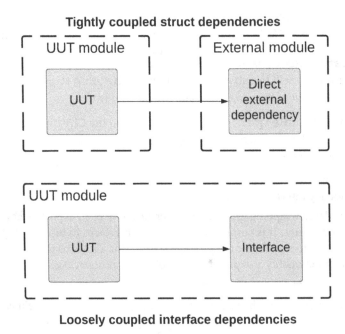

Figure 3.3 – Creating loosely coupled code using interfaces

As we can see, representing dependencies by using internally defined interfaces allows us to break the hard dependencies between modules. The underlying external `struct` that satisfies this interface can be created and injected into the UUT outside this package, without introducing a hard dependency.

Furthermore, since structs can satisfy multiple interfaces, they also give us the flexibility to reduce the scope of operations that we want to have access to inside the UUT. This is particularly useful when working with large SDKs or complex external APIs, where we might not want to define or mock all the functions.

Mocking frameworks

Now that we have refactored the code to leverage the power of interfaces and create loosely coupled code, let us see how we can make use of their power during testing as well.

When it comes to creating mocks, two popular mocking frameworks allow us to easily generate and assert on mocks:

- `golang/mock` is an open source framework that was first released in March 2011. You can read all about it at `https://github.com/golang/mock`. It consists of a mocking package and a code generation tool, `mockgen`.

- `testify/mock` is an open source framework that was released in October 2012. You can read all about it at `https://github.com/stretchr/testify/#mock-package`. Just like `golang/mock`, it consists of a mocking package and a code generation tool, `mockery`.

The two frameworks offer a lot of similar functionality, so choosing one can seem like a bit of an arbitrary choice. At the time of writing, the `testify/mock` package has been imported by over 13,000 packages (see `https://pkg.go.dev/github.com/stretchr/testify/mock?tab=importedby`), while the `golang/mock` package has been imported by over 12,000 packages (see `https://pkg.go.dev/github.com/golang/mock/gomock?tab=importedby`). This further underlines that they are two very popular frameworks for Go developers.

As we will see in the next section, *Working with assertion frameworks*, `testify` also provides a very powerful and popular assertion framework. Therefore, we will use `testify/mock` as our mocking solution throughout this book.

To use this framework, you will need to install its two main components by running these commands, which are correct at the time of writing:

```
$ go get github.com/stretchr/testify
$ go install github.com/vektra/mockery/v2@latest
```

These two commands will set up the framework for us to use going forward. Make sure you run these two commands to be able to follow along with the code examples provided throughout this book. While we will use this framework for mocking, the concepts discussed apply to `golang/mock` as well.

Generating mocks

So far, we have prepared our dependencies, selected a mocking framework, and then installed it. Now, let us learn how to put it to use. We previously mentioned that `testify` provides a code-generation tool for creating mocks. This tool makes it easy to generate boilerplate mock code so that we do not need to create and maintain it by hand.

Mock generation in `testify` does not require any special annotations. Mocks can be generated for interfaces and functions, making them suitable for both **function substitution** and **interface substitution**.

The `mockery` command has support for a variety of flags. Here are some common ones you might see:

- The `--dir string` flag specifies the directory in which to look for interfaces to mock.
- The `--all` flag specifies to search for through all subdirectories and generate mocks.
- The `--name string` flag specifies the name or regular expression to match while searching for interfaces to generate mocks.
- The `--output string` flag specifies the directory to place generated mocks into. By default, this is configured to be `/mocks`.

You can see all the other options available for this command by using `mockery -help`.

We can now generate mocks for our interfaces by using the following command:

```
$ mockery --dir "chapter03" --output "chapter03/mocks" --all
```

This command looks for all interfaces in the `chapter03` directory and all its subdirectories and places the generated files in the `chapter03/mocks` directory. The output of this command should look like this:

```
11 Sep 22 17:38 BST INF Starting mockery dry-run=false
version=v2.14.0
11 Sep 22 17:38 BST INF Walking dry-run=false version=v2.14.0
11 Sep 22 17:38 BST INF Generating mock dry-run=false
interface=OperationProcessor qualified-name=github.com/
PacktPublishing/Test-Driven-Development-in-Go/chapter03/input
version=v2.14.0
11 Sep 22 17:38 BST INF Generating mock dry-run=false
interface=ValidationHelper qualified-name=github.com/
PacktPublishing/Test-Driven-Development-in-Go/chapter03/input
version=v2.14.0
```

As we can see from the output, our two interfaces, `OperationProcessor` and `ValidationHelper`, have been detected and mocks have been generated for them. The generated files will contain structs that will satisfy the defined interfaces:

```
// OperationProcessor is an autogenerated mock type for the
// OperationProcessor type
  type OperationProcessor struct {
  mock.Mock
}


// ProcessOperation provides a mock function with given
```

```
// fields: operation
func (_m *OperationProcessor) ProcessOperation(operation
calculator.Operation) (*string, error) {
  ret :=_m.Called(operation)
// implementation code
}
```

The generated structs also contain a nested struct of the `mock.Mock` type. This provides functionality for asserting activity on the mock. This functionality is important when verifying mocks, which we will explore next.

> **Regenerating mocks**
>
> It is common for engineering teams to add the mock generation to the specification of their Docker files. This will allow the mocks to be generated as part of the CI/CD pipeline and be used during the build process.

Verifying mocks

We are now ready to begin writing tests for the `Parser` struct, which uses the generated mocks that we've created. *Figure 3.4* depicts the steps that we follow to write the tests:

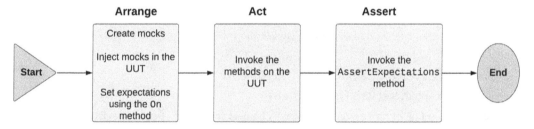

Figure 3.4 – Writing tests using mocks

The rough procedure of how our tests will look like using the generated mocks is as follows:

1. **Create mocks**: We create the mock structs in the **Arrange** step of our test. The mock will be separated from any transitive dependencies, so it will be easy to initialize. At this point, we should have as many mocks as there are direct dependencies of the UUT.

2. **Inject mocks in the UUT**: We inject the mock while creating the UUT in the **Arrange** step of our test. Since the mocks satisfy the interfaces of the real dependencies, the UUT is not aware of whether it is receiving a real dependency or a mock.

3. **Set expectations using the** On **method**: We invoke the On method on the mock to set up any expectations of the mock behavior. We also set up any expected parameter invocations and return values. This concludes the **Arrange** step of your test.

4. **Invoke the methods on the UUT**: We write the **Act** section of our test as normal. The UUT is unaware that it will be using a mock during its operation, so any method invocations will function as normal.

5. **Invoke the** AssertExpectations **method**: Finally, in the **Assert** section of our test, invoke the AssertExpectations method on all of the mocks to ensure that all the previously declared expectations are verified.

The usage of mocks is very simple and integrates well with the testing library. Let us have a look at a simple test of the Parser type:

```go
func TestProcessExpression(t *testing.T) {
    t.Run("valid input", func(t *testing.T) {
        // Arrange
        expr := "2 + 3"
        operator := "+"
        operands := []float64{2.0, 3.0}
        expectedResult := "2 + 3 = 5.5"
        engine := mocks.NewOperationProcessor(t)
        validator := mocks.NewValidationHelper(t)
        parser := input.NewParser(engine, validator)

        validator.On("CheckInput", operator,
            operands).Return(nil).Once()
        engine.On("ProcessOperation", &calculator.Operation{
            Expression: expr,
            Operator: operator,
            Operands: operands,
        }).Return(expectedResult).Once()

        // Act
        result,err := parser.ProcessExpression(expr)

        // Assert
```

```
    // other assertions
    validator.AssertExpectations(t)
    engine.AssertExpectations(t)
    })
}
```

The simple `TestProcessExpression` highlights the usage of mocks in test writing. The usage of the On method allows us to easily configure expected behavior for all mocked direct dependencies.

As demonstrated, the On method can be used to specify detailed expectations. Here are some of the ones you will often encounter:

- The **function name** is specified during the call of the On method itself. The first parameter of the On method is the name of the function that should be mocked.

- The **function parameters** are also specified as parameters to the On method. The arguments can be specific values or we can assert their type using the `mock.AnythingOfType` function. We can also use `mock.Anything` if we don't care about making any validation of the given argument, but this should be used sparingly because the intention behind the test might be hard to understand later.

- The **return values** are specified with the chained `Return` method, which is invoked after the On method. It allows you to return specific values if the specified method is invoked with the configured function parameters.

- The **invocation count** is also specified using chained methods after the On method. There are shorthand methods for `Once` and `Twice`; otherwise, the `Times` method can be used to specify a custom invocation count. The `Unset` method can be used to specify that a mock handler should not be called.

> **Verifying expectations**
>
> Remember to call the `AssertExpectations` method on each of your mocks if you want to assert that they have been called according to the expectations laid out by your On methods. This gives you fine-grained control over how the UUT interacts with its dependencies.

Consider how much code you would have to write to set up custom types according to preconditions and then also verify that dependencies have been invoked according to expectations. The `testify/ mock` library makes it easy for us to leverage the power of mocks in a unified way across all our projects.

Working with assertion frameworks

While the `testify/mock` functionality is useful for creating mocks, `testify` is best known for its assertion framework. In this section, we will explore some common assertion frameworks and how we can use them to further streamline and expand our tests.

So far, we have been writing our verifications using `if` statements and invoking the correct failure method on the `testing.T` parameter:

```
// Assert
if err != nil {
  t.Fatal(err)
}
```

This approach is simple, but it does have the following disadvantages:

- **Repetition**: A lengthy or complex test will end up making multiple assertions. We will then have to repeat this error assertion block multiple times, making the test verbose.

- **Difficult to make advanced assertions**: We want to have the same fine-grained control over the verifications we undertake on our mocks throughout the rest of the test.

- **Completely different approach to other languages**: This approach is completely different from other programming languages, which have powerful mocking and assertion frameworks. Java's `JUnit` is such an example.

While the Go standard library does not provide functionality for assertions, two popular assertion frameworks provide this functionality:

- `testify` is an open source assertion framework that provides an easy-to-use and powerful assertion package. The `assert` package provides this functionality. You can read about it at `https://github.com/stretchr/testify#assert-package`.

- `ginkgo` is an open source assertion framework that provides **behavior-driven development (BDD)** style test writing and assertions. You can read about it at `https://github.com/onsi/ginkgo`. Adopting this style of testing allows developers to write tests that read like natural language.

We will discuss BDD style tests in *Chapter 5, Performing Integration Testing*. Therefore, we will reserve the discussion of writing this type of test until then. We will continue our current exploration with the `testify` framework.

Using testify

The `assert` package provides many useful functions for creating fine-grained assertions. Here are some of the ones that you will encounter often:

- **Equality assertions**: The `assert.Equal` function allows you to check whether two objects are equal. If the types checked are pointer-based, a value check on the reference values will be conducted. The opposite function, `assert.NotEqual`, also exists:

```
assert.Equal(t, expected, actual)
assert.NotEqual(t, expected, actual)
```

- **Nil assertions**: The `assert.Equal` function should not be used for nil values. Instead, the `assert.Nil` method should be used. The opposite function, `assert.NotNil`, also exists:

```
assert.Nil(t, actual)
assert.NotNil(t, actual)
```

- **Contains assertions**: The `assert.Contains` function verifies that a specified value is contained inside a string, list, or map. The opposite function, `assert.NotContains`, also exists:

```
assert.Contains(t, collection, element)
assert.NotContains(t, collection, element)
```

- **Subset assertions**: The `assert.Subset` function verifies that all the values in a specified subset are contained in a specified list. The opposite function, `assert.NotSubset`, also exists:

```
assert.Subset(t, list, subset)
assert.NotSubset(t, list, subset)
```

The `testify/require` package also provides the same assertions, but will terminate the test in the case that an assertion fails. This package should be used in the case of fatal test errors.

For example, we can replace our previous `if` statement, which makes a call to `t.Fatal`, with the following single line of code:

```
require.Nil(t, err)
```

> **Augmenting the testing package**
>
> You should use assertion frameworks to complement the simplicity of the `testing` package. As you begin to write more Go code, you should familiarize yourself with the assertion frameworks functionality and begin to use them in your tests.

Asserting errors

One final aspect to cover when discussing assertions is how to verify errors. Sometimes, we want to verify not only that an error occurs, but that the correct error message is also returned. You should ensure that your tests verify such error messages when appropriate.

The `assert.EqualError` function verifies that a returned error is not `nil` and that its message is equal to the provided string. This will make it easy to verify your error messages. As with all the functions we have seen so far, the `require` package also provides this function.

Let us look at an example that verifies an error scenario:

```go
t.Run("invalid operation", func(t *testing.T) {
    // Arrange
    expr := "2 % 3"
    operator := "%"
    operands := []float64{2.0, 3.0}
    expectedErrMsg := "bad operator"
    engine := mocks.NewOperationProcessor(t)
    validator := mocks.NewValidationHelper(t)
    parser := input.NewParser(engine, validator)
    validator.On("CheckInput", operator, operands).
        Return(fmt.Errorf(expectedErrMsg)).Once()

    // Act
    result, err := parser.ProcessExpression(expr)

    // Assert

    require.NotNil(t, err)
    require.Nil(t, result)
    assert.Contains(t, err.Error(), expr)
    assert.Contains(t, err.Error(), expectedErrMsg)
    validator.AssertExpectations(t)
})
```

The test creates a variable called `expectedErrMsg`, which represents the error message that the mock will return. This message is then passed to the `assert.Contains` function, which will verify it against the error returned by the `ProcessExpression` method invoked on the UUT.

> **Custom error types**
>
> You can also create your own custom error types instead of relying solely on Go's built-in `error` type. This will provide you with type safety on error checking, instead of relying on error messages, which might change and make your tests brittle.

Mocks and assertion frameworks are tools that we use to easily write tests. However, even the most skilled test writer will struggle to test code that is badly designed. The iterative nature of TDD together with good software design principles will result in testable, maintainable code.

Writing testable code

The final aspect we will cover in this chapter is how to write testable code using the **SOLID** software design principles. As we have seen multiple times already, well-designed code is also easy-to-test code. Application code that is difficult to test is often a sign that the application will be hard to change and maintain.

These five powerful principles were introduced in a paper by Robert C. Martin in 2000, then later published in his book *Agile Software Development, Principles, Patterns, and Practices*. These principles help Agile teams deliver maintainable, easy-to-refactor code.

Figure 3.5 summarizes the SOLID design principles:

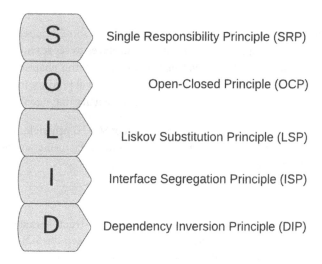

Figure 3.5 – The SOLID design principles

Let us recap the SOLID principles and what they mean for test writing:

1. **Single Responsibility Principle (SRP)**: For this, entities should have a single job and a single reason to change. This principle will keep testing code simple since the scope of the functionality provided by the class is small. We can then focus our efforts on covering edge cases, as opposed to covering a large number of methods.

2. **Open-Closed Principle (OCP)**: For this, entities should be open for extension, but closed for modification. This principle translates to ensuring that code changes extend but don't break existing behavior. Code that is backward-compatible by design will not require numerous test changes. Instead, the new/extended functionality can be covered by new test cases, ensuring that the test suite continues to be stable.

3. **Liskov Substitution Principle (LSP)**: For this, every subclass or derived class should satisfy the behavior of its parent or base class. Since Go does not have inheritance, you might be tempted to conclude that it does not apply. However, we achieve polymorphism using interfaces, so we can express this principle in terms of the contracts they set. Code that maintains substitutable interfaces will be easy to test, as it will again not require many test changes to the existing suite.

4. **Interface Segregation Principle (ISP)**: For this, client code should not be forced to implement methods or interfaces that it does not use. This principle encourages the usage of small interfaces, which only wrap around a single responsibility. Interfaces should be defined on the side of the client/calling code, which should only define interface methods for the functionality they are interested in using. Small interfaces lead to small mocks, which, in turn, lead to simple test setups and assertions.

5. **Dependency Inversion Principle (DIP)**: For this, entities should depend on abstractions, not concretions. This principle encourages using interfaces to represent dependencies. We discussed this principle in the *Dependency injection* section of this chapter. We saw that using this technique in combination with the power of interfaces will produce loosely coupled code that is easier to test, due to fewer out-of-package dependencies that need to be set up.

As we have seen, interfaces are central to the implementation of SOLID principles in Go. They should be used to write code that is easy to maintain and refactor. Since refactoring is a central part of the TDD practice, easy-to-refactor code will also be easy to test code.

> **The SOLID entity**
>
> Remember that in Go the entity should be the package, not the struct. Packages provide their own APIs, which establish their own contracts and interfaces. Keep the SOLID principles in mind when designing your package APIs.

Summary

In this chapter, we expanded on unit testing essentials by learning how to handle code dependencies. We began by introducing DI and exploring different approaches to how this is done in Go. You learned how to use mocks to have fine-grained control over the preconditions that your test runs in, as well as how to generate them using the `testify/mock` framework. Then, we explored different assertion frameworks and how to use them to streamline our tests. Finally, we discussed the SOLID principles, which will help us to write testable code.

In *Chapter 4, Building Efficient Test Suites*, we will begin to look at our tests as a collection, ensuring they complement each other and cover edge cases. We will also explore the popular Go technique of **table-driven testing**.

Questions

Answer the following questions to test your knowledge of this chapter:

1. Continue the existing implementation of the simple calculator by implementing and testing the remaining operations:

    *   Subtraction

    *   Multiplication

    *   Division

2. Ensure that you use all the techniques we have used so far to produce well-tested code.

Further reading

To learn more about the topics that were covered in this chapter, take a look at the following resources:

*   *Clean Architecture: A Craftsman's Guide to Software Structure and Design: A Craftsman's Guide to Software Structure and Design*, Robert C. Martin, published by Addison-Wesley

*   *Design patterns: elements of reusable object-oriented software*, Erich Gamma et al., published by Addison Wesley

*   *Compile-time Dependency Injection With Go Cloud's Wire, The Go blog*, available at `https://go.dev/blog/wire`

4

Building Efficient Test Suites

In the previous chapter, we learned how to supplement the functionality of Go's testing package with third-party libraries. These libraries make it easier to mock the dependencies of the **Unit Under Test (UUT)** and create assertions in these tests. Mocks are essential building blocks to being able to easily write test code for well-designed implementation code, according to the **SOLID** design principles.

In practice, developers identify edge cases of their requirements and implementations, ensuring a good **code coverage** percentage, which we discussed in *Chapter 2, Unit Testing Essentials*. In this chapter, we will learn how to create test suites.

One popular technique for constructing test suites in Go is **table-driven testing**. We will learn how to build tables that cover edge cases and exercise the UUT with a variety of inputs, ensuring that the UUT has a stable implementation. We will also leverage some of the techniques we've explored so far, such as test setup, subtests, and mocks.

One of the most popular usages of Go is to build web applications, and in this chapter, we will explore how to build and test just that. We will move on from the simple calculator example that we have looked at so far and look at a new use case: the BookSwap application. This service will allow users to create book listings they wish to swap, allowing others to borrow them.

This example will involve building a REST API with Go's net/http package and learning how to test it. It is particularly important to cover edge cases when dealing with user input, so we will test the BookSwap API using the techniques covered so far.

In this chapter, we will cover the following main topics:

- What edge cases are and how to identify them
- How to test web applications and APIs, which may rely on external services
- The popular Go testing technique of table-driven testing
- The BookSwap use case application

Technical requirements

You will need to have **Go version 1.19** or later installed to run the code samples in this chapter. The installation process is described in the official Go documentation at `https://go.dev/doc/install`.

The code examples included in this book are publicly available at `https://github.com/PacktPublishing/Test-Driven-Development-in-Go/chapter04`.

Testing multiple conditions

So far, we have covered how to structure and write tests. However, developers need to know *what* aspects of their code to test, as well as *how* to test them. Remember that the lower we go on the **testing pyramid**, the cheaper and faster the tests are to run. Therefore, it is important for developers to know how to exercise their code as low in their stack as possible. In this chapter, we'll focus on covering edge cases as part of our developer **testing strategy**.

As discussed in *Chapter 1, Getting to Grips with Test-Driven Development*, automated tests should be based on the system requirements we implement. In general, system requirements will focus on the specification of the success scenarios and system functionality additions. Designing your testing strategy around these requirements serves the primary purpose of ensuring that your system satisfies its functional requirements.

A secondary purpose of your testing strategy should be to verify the behavior and robustness of your system-under-failure cases, such as incorrect/unexpected input, transient errors, or slow responses. Developers need to ensure that their systems are able to gracefully handle all sorts of operating conditions.

We will learn how to identify these conditions and apply the testing techniques we've learned so far to devise testing strategies that give us confidence in our solutions, regardless of the inputs and the conditions our system operates in.

Figure 4.1 shows the dual nature of testing, comprising positive and negative tests, to ensure that both functionality and error handling is correctly implemented in our systems:

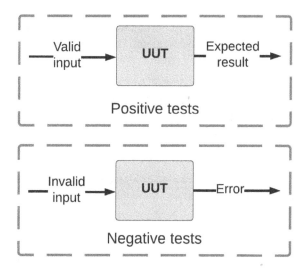

Figure 4.1 – The two types of tests

The two types of tests are as follows:

- **Positive tests**, which use valid input for the UUT and verify that the expected result is returned from the UUT. This type of test ensures that the application behaves correctly according to the functional requirements. Positive tests cover the following:

  - How the UUT handles valid input

  - How the UUT behaves in expected scenarios

  - How the UUT satisfies system requirements

- **Negative tests**, which use invalid input for the UUT and verify that an error is returned from the UUT. This type of test ensures that the application can gracefully handle invalid input, with meaningful errors and avoiding crashes. Negative tests cover the following:

  - How the UUT handles invalid input

  - How the UUT behaves in unexpected scenarios

  - How the UUT behaves outside of system requirements

Each of these tests is comprised of different types of test scenarios of varying complexity, based on the values of input variables and their combinations.

> **The importance of negative tests**
>
> Both positive and negative tests are equally important for production systems. Error handling is an important part of the user journey. We want users to receive meaningful messages in the case of errors, as well as recover successfully in the case of slowdowns or outages.

Happy path testing or **happy flow testing** is the verification of the default success scenario without any errors or exceptions. Covering the default and requirement-specific scenarios ensures that the system behaves well in ideal scenarios. However, as developers, we need to know more than the ideal behavior of our systems.

Figure 4.2 depicts the different types of test cases for a given input parameter of our system. The different types of test cases cover the entire range of possible input parameter values:

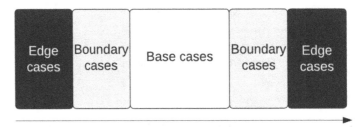

Figure 4.2 – The types of test cases of a given input variable

A good testing strategy should cover the following four major types of test cases of a given input variable:

- **Base cases** occur at the expected values of an operating parameter. For example, given an input parameter representing a name, a base case for it would be a short valid string value. These cases are often defined in the system requirements and make up the scenarios of the happy path testing strategy.

- **Edge cases** occur at the extreme of an operating parameter. For example, given a string input parameter, some edge cases for it would be an empty string value, a multiline string, or a string with special characters.

- **Boundary cases** occur on either side of an edge case, approaching the extreme values of an operating parameter. These cases are particularly important for asserting values that must have a particular value. For example, given a numeric input parameter representing temperature for a water temperature measurement application, we could boundary-test its values around the water freezing point and water boiling point.

As demonstrated by these examples, edge cases are often based on the data type of the input/user parameter, as well as their purpose. We will explore other types of parameters and how to identify their extreme/edge-case values in the next section.

Systems will often operate on multiple input variables. The combinations of input variables and their edge cases can result in different system behavior.

Figure 4.3 demonstrates the final type of test case, which tests the particular scenario of multiple edge cases of input variables:

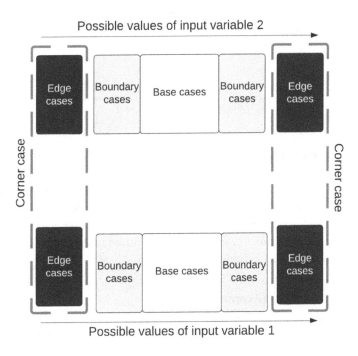

Figure 4.3 – Corner cases

Corner cases occur at extremes or edge cases of multiple operating parameters. Any combination of edge cases between the two types of input variables would result in a corner case. For example, given multiple string input parameters, we would achieve a corner case by a combination of any of the edge cases of these parameters.

Figure 4.4 demonstrates the test case combinations of two input variables:

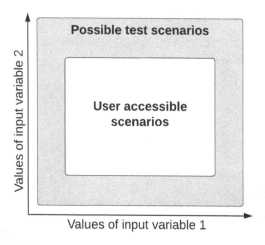

Figure 4.4 – Combining two input variables in a testing strategy

As the number of input parameters of our system increases, the number of combinations of edge cases increases, resulting in a large number of corner cases that must be tested. In order to minimize test writing and maintenance efforts, it's important to identify the subset of user-accessible scenarios from the entirety of possible test scenarios. These should be prioritized in the testing strategy, and testing can then be extended as the project becomes more mature.

> **The difference between edge and corner cases**
>
> The terms *edge case* and *corner case* are often used interchangeably. One easy way to remember the difference is that an edge case pushes the extremes of a parameter, while a corner case combines these extremes by pushing the user to a corner configuration.

Identifying edge cases

There is no particular well-defined procedure for identifying edge cases for variables and algorithms. This is where the experience of software testers and engineers makes a big difference, as they can intuitively identify edge cases of code and requirements upon inspection. We can, however, make some recommendations of what to watch out for.

Figure 4.5 demonstrates special cases based on the variable type:

String type	Numeric type	Custom `struct` type	Collection types
• Empty string • Long string • Special characters • Multi-line string	• Zero value • Min/Max value according to numeric type • Positive and negative values	• Zero value • Nil value (if referenced by pointer) • Combinations of set fields	• Zero elements • One element • Nil value • Duplicate elements • Large amount of elements

Figure 4.5 – Special cases of different variable types

The special cases of variable types are as follows:

- String-type variables have the following special cases:

  - An empty or zero character string—`" "`.

  - A long string, which exceeds the expected length of the base-case valid string—`"a very very very very long string"`.

  - A string containing special characters, including Unicode characters and special accent characters—`"a $p€¢iał string!"`.

  - A multiline string containing new line delimiters—`"a multi \n line string"`. Remember that Go allows the definition of raw string literals by the use of backticks, which can also contain other special characters.

- Numeric-type variables have the following special cases:

  - A zero value—`0`.

  - The minimum and maximum values are according to the numeric type. For example, the `int8` type has a minimum value of `-128` and a maximum value of `127`, while the `uint8` type has a minimum value of `0` and a maximum value of `255`. These values increase according to the memory allocation of the given type.

  - Positive and negative numeric values may also require special handling, according to the logic of the UUT.

- Custom `struct` types have the following special cases:

    - The zero value of the custom struct, with no initialization—`a := MyType{}`.

    - The nil value of the type, if passed by a pointer—`var a *MyType`.

    - Combinations of initialized and uninitialized fields of the given type—`a := MyType{ field1: "Value"}`. Testing these combinations can reveal whether any fields should be added to initialization/constructor functions. While Go does not provide default implementations of constructors, it is common to declare package - scoped functions that initialize an instance and return it—`func NewMyType(v string) *MyType`.

- Collection types wrap around Go's in-built collection types—arrays, slices, and maps:

    - Zero-element or empty collection—`c := []int{}`.

    - One-element or single-element collection—`c := []int{0}`.

    - Nil value or collection with no allocated memory—`var c []int`.

    - Duplicate elements—`c := []int{0, 0}`.

    - Collection with a large number of elements—`var c [999]int`.

The special cases of each variable type should inform your decision as to which edge cases you should attempt to cover, but you should extend your edge cases to cover the boundaries of any system requirements and edge cases.

When formulating your test cases, you should break down the UUT into small logical blocks, identify their inputs and their edge cases, then construct your test suites to verify these cases accordingly. We will learn how to easily write test suites using table-driven testing later on in this chapter.

External services

Now that we understand how to identify the edge cases of input parameters based on their type and system requirements, we can now turn our attention to testing with external services. As discussed in *Chapter 3, Mocking and Assertion Frameworks*, any direct dependencies of the UUT should be mocked, allowing us to test the UUT in isolation.

As the Go package provides us with an easy way to build small, self-contained APIs, we can treat all dependencies as external services. These dependencies can be divided into two categories:

- **Internal system dependencies** are located inside the system we are testing, whether inside the same service or not. We have full control of these dependencies.

- **External system dependencies** are located outside the system we are testing, providing extra functionality such as a database or third-party functionality. We do not have full control of these dependencies.

Always mock external system dependencies

As we do not control system dependencies, testing against their live/real versions could introduce brittleness and extra costs to our test suites. With the exception of databases, you should always mock your external system dependencies. We will explore database testing further in *Chapter 5, Performing Integration Testing*.

When it comes to edge cases of external system dependencies, these APIs will most often connect with our system using some kind of network connection. Their edge cases are heavily influenced by this connection.

Figure 4.6 depicts the possible errors that can happen in the integration between the UUT and the external service:

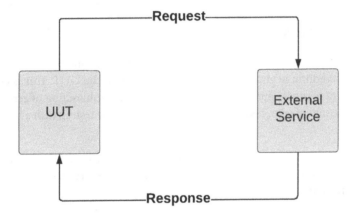

Figure 4.6 – Possible errors in communication between the UUT and external service

When each part of the exchange is happening over a network, both the request and response need to be tolerant of delays and retries:

- The external service may error out and respond with an internal service error. In this case, the UUT will need to handle a full outage and serve a default response.

- The request may take a long time to be delivered to the external service. In this case, the UUT will need to wait for the response for a pre-defined amount of time and then consider the request failed. The UUT may then decide to retry this request to fetch any resources.

- The response from the external service may not arrive at all. In this case, the UUT will need to retry the whole request cycle and handle this duplicate flow in the application logic.

Errors are an inevitable part of writing code and running applications, especially those relying on external services for their functionality.

Modern systems will rely on many types of external APIs, which can communicate over REST APIs, RPC calls, or even asynchronously through event buses. The testing concerns of these integrations are similar, as the communication between the UUT and the external service will be the most error-prone part of the integration.

> **Idempotence as design**
>
> In API design, idempotent operations can be called repeatedly without changing the initial result. It is considered good practice to design all operations as idempotent to ensure that operations can be retried in the case of error recovery.

Error-handling refresher

Up until now, we've discussed how to identify possible edge cases and write tests for them, but resilience and error handling begins with the implementation of the UUT. This is especially true for Go developers, where the language design requires the explicit handling of error cases. Let us supplement our discussion of edge case identification and error case testing with a brief summary of error handling in Go.

Error handling plays a vital role in writing Go code. The Go team has opted for explicit error handling with the built-in `error` type, in order to avoid exceptions and `try-catch-finally` style blocks, which can cause brittle and error-prone code.

The `error` type is a simple interface:

```
type error interface {
    Error() string
}
```

This interface also makes it easy for us to create our own custom error types, which just need to implement the `Error() string` method. Errors are returned just as any other value, most commonly using **multiple return values**, and they are handled just like any other return value.

For example, we've already seen the `Parser` calculator return an error, in the case of an invalid mathematical expression:

```
func (p *Parser) ProcessExpression(expr string) error
```

The zero value of the error type is `nil`. Most commonly, a `nil` error value will signal that no issues have occurred during execution.

It is common practice to handle errors first in the code by calling the possibly failing expression:

```
if err := parser.ProcessExpression(*expr); err != nil {
   log.Fatal(err)
}
```

In this example, we initialize the `err` variable at the same time as the call to the erroring function, limiting the scope of the variable to the `if` statement block.

Note that we check for the presence of an error, not its absence. In the case of `err != nil`, we simply kill the application with a call to the `log.Fatal` function. This is the typical way we handle errors in Go.

Handling errors explicitly with the `error` type has the following advantages:

- **Guarantees that error cases are handled, avoiding any later panics or nil pointers**: Handling errors first, at the top of the function code, reduces checks for valid data later on in the function code. This can help simplify code execution flow.

- **Makes it easy to see which error scenarios we need to cover in our testing strategy**: The function signature will show which methods and functions can produce errors, forcing calling code to handle them explicitly.

- **Gives us a unified way of representing error states and returning error messages**: The built-in `error` type gives all Go codebases a unified way of representing error states, which makes it easy to construct and return user-facing errors.

However, some developers find the error-checking code blocks repetitive and verbose. A common criticism is that they need to handle all errors, even those that are relatively unlikely to happen. Error return values may be disregarded by using the **blank identifier** (the _ operator) or not assigning the return value to any variable, but this is commonly discouraged.

You can make your own opinions on Go's explicit error handling, but we will use it throughout this book as it is a convention and standard practice for how we write Go.

> **Handling errors first but returning them as the last parameter**
> In a function with multiple return values, remember that the error type is typically the last return value. You should then handle the error case first, returning in the case of abnormal scenarios, and keeping your code minimally indented.

Table-driven testing in action

Now that we have discussed the fundamentals of identifying edge cases and handling errors, we can begin to look at how to build test suites that cover a variety of scenarios. A popular technique in Go is to use **table-driven testing**. This technique uses the fundamentals we've learned so far to structure test suites that cover a variety of scenarios.

Let us begin with a simple example to demonstrate the test-writing process. We will implement a new `Divide` mathematical operation that does the following:

- Returns the result formatted as a string to two decimal points
- Returns an error in the case that the divisor is 0

From the preceding requirement, we can formulate the following signature for this new operation:

```
func Divide(x, y int8) (*string, error)
```

We remember that the minimum value of `int8` is `-128` and the maximum value is `127`.

As previously discussed, we make use of multiple return values to encourage explicit error handling in the calling code. Based on the functionality requirements and the lessons learned from the previous *Identifying edge cases* section, we can identify the following test cases:

- **Base cases**:
  - Two positive values for x and y
  - Two negative values for x and y
- **Edge cases**:
  - Equal values for x and y
  - Maximum value for x and positive value for y
  - Minimum value for x and positive value for y
  - A zero value of x and a nonzero value for y
  - A positive value of x and a zero value for y
- **Corner cases**:
  - Zero values for x and y
  - Maximum values for x and y
  - Minimum values for x and y

In *Chapter 2, Unit Testing Essentials*, we saw how to write tests and implement different scenarios with subtests. This involves declaring a shared test setup and declaring a subtest for each case. For example, the implementation of the first test case could look like this:

```go
func TestDivide(t *testing.T) {
  t.Run("positive x, positive y", func(t *testing.T) {
    x, y := int8(8), int8(4)
    r, err := table.Divide(x, y)
    assert.Nil(t, err)
    assert.Equal(t, "2.00", *r)
  })
}
```

As we can see from the highlighted lines in this code snippet, the following components are the ones that change according to the test case we are running:

- The name of the test case, which will make our test output easy to read

- The inputs that will change values according to the test case we are running

- The expected result value and error value according to the test case we are running

As can be seen from the previous snippet, there is quite a bit of boilerplate code that can be reused across test cases:

- The declaration of the test function and any required UUT setup

- The declaration of the subtest and its nested testing function

- The invocation of the `Divide` function with its input values

As the interaction with the `*testing.T` object is the most verbose part of the test implementation, a shorter and simpler alternative to test cases is to use table-driven tests, which we will learn all about in the next section.

Implementing table-driven tests has a very simple recipe. We will use the example of the `Divide` function from the previous section to demonstrate each step.

Step 1 – declaring the function signature

We begin by declaring the function signature that we have presented previously and writing only enough code to make the code compile:

```
package table

func Divide (x, y int8) (*string, error) {
    return nil, nil
}
```

The signature of the function returns a pointer to a string and an error. In practice, we expect only one of the two values to ever be `nil`:

- In the normal flow, the result string will be non-nil, and the error value will be nil

- In the abnormal flow, the result string will be nil, and the error value will be non-nil

Therefore, by setting both values to nil, we will guarantee that there will be no accidentally passing test cases. This helps us begin the red phase of the **red-green-refactor test-driven development (TDD)** process.

Step 2 – declaring a structure for our test case

The first step of writing test code is to declare a custom type to wrap around our test case. The purpose of this structure is to hold the inputs and expected outputs of the test case. Generally, this type is declared inside the scope of the function test, but it can also be shared across tests.

The test case of our `Divide` function looks like this:

```
func TestDivide(t *testing.T) {
  type testCase struct {
    x, y     int8
    wantErr error
    want     *string
  }
}
```

This custom type is a simple `struct` type that wraps around x and y—the two inputs of the function and the two expected results of the function—the formatted result and the possible returned error. Note that in Go it is customary to name the expected result as want or with the want prefix. This is different from other languages, where the naming convention begins with the word `expected`.

Step 3 – creating our test-case collection

Now that we have a way to express our test cases, we can begin to create a collection of all the cases we want to test for our function. Based on the two base cases that we identified for the Divide function in the previous section, we can create the following tests collection:

```
tests := map[string]testCase{
    "pos x, pos y":    {x: 8, y: 4, want: "2.00"},
    "neg x, neg y":    {x: -4, y: -8, want: "0.50"},
}
```

We prefer to use map to add a corresponding name to the test case, which allows us to add the name as a key and the test case as a value. An alternative solution is to use a slice and save the name of the test case as a field in the testCase type.

Note that we don't provide a value to the wantErr field in the previous test case, as the base cases do not require the verification of errors. The zero value of the error type is nil, so not setting a value for it will be equivalent to declaring a happy path test case.

We can further optimize our implementation of the tests map by using anonymous struct types for our testCase type to reduce boilerplate and keep the scope of the testCase type small:

```
tests := map[string]struct {
    x, y int
    wantErr error
    want string
}{
    "pos x, pos y": {x: 8, y: 4, want: "2.00"},
    "neg x, neg y": {x: -4, y: -8, want: "0.50"},
}
```

This can further shorten the test declaration but will not allow us to share the testCase type between tests.

Step 4 – executing each test

With our table of tests in place, we will execute each test case as a subtest. We will use the `range` statement to loop through the map of tests, which will return the name of the test case and the test case instance itself. Then, we pass the test name as the subtest name and use the test case during the test setup and execution:

```
for name, tc := range tests {
    t.Run(name, func(t *testing.T) {
        // Test execution
    })
}
```

This step allows us to set up the interaction with the test runner in one single block for the entire test suite. Remember that each subtest is its own function, so we can individually fail tests or stop the execution of the entire test suite using the `testing.T` helpers that we have explored in *Chapter 2, Unit Testing Essentials*.

Step 5 – implementing the test assertions

Once we have set up the `tests` map and its interaction with the test runner, we can begin to implement the testing logic based on the inputs and outputs defined in the `testCase` type:

```
for name, tc := range tests {
    t.Run(name, func(t *testing.T) {
        x, y := int8(tc.x), int8(tc.y)
        r, err := table.Divide(x, y)
        if tc.wantErr != nil {
            assert.Equal(t, tc.wantErr, err)
            return
        }
        assert.Nil(t, err)
        assert.Equal(t, tc.want, *r)
    })
}
```

Based on the `tc` test case value retrieved from the `tests` map, we use its values of x and y to invoke the `Divide` function. Then, we verify the error value and the result value from the `tc` test case as well. Note that, just as we do with error handling, we verify the error value first and return from the test in the case of the error case.

Step 6 – running the failing test

Our table-driven test suite has successfully been implemented in five easy steps! The basics of running tests and assertions are in place, so we can now run the tests and see them fail. We can now run the test with the go test command, as we have done so far:

```
$  go test -run TestDivide ./chapter04/table -v
--- FAIL: TestDivide (0.00s)
  --- FAIL: TestDivide/pos_x,_pos_y (0.00s)
```

As we can see from the output, all the tests are run in their own subtest with the given scenario name passed to the test runner. The -v flag is the verbose flag, which will show the full output of all the tests that are run.

Step 7 – implementing the base cases

We now begin to implement the Divide function happy path cases. We will write two simple lines of code that will allow the tests of the base cases to pass:

```
func Divide(x, y int8) (*string, error) {
  r := float64(x) / float64(y)
  result := fmt.Sprintf("%.2f", r)
  return &result, nil
}
```

These two lines of code will handle the normal program flow. We then rerun the base case tests that we have written so far and see them pass:

```
$  go test -run TestDivide ./chapter04/table -v
--- PASS: TestDivide (0.00s)
  --- PASS: TestDivide/pos_x,_pos_y (0.00s)
  --- PASS: TestDivide/neg_x,_neg_y (0.00s)
```

Once these tests pass, we enter the green phase of the red-green-refactor TDD process.

Step 8 – expanding the test case collection

With the base case tests passing, it's time to expand our test case collection to include error cases. Based on the 10 test cases that we identified for the `Divide` function in the previous section, we can add the following cases to the `tests` collection:

```
tests := map[string]struct {
  x, y int
  wantErr error
  want string
}{
  "pos x, pos y":   {x: 8, y: 4, want: "2.00"},
  "neg x, neg y":   {x: -4, y: -8, want: "0.50"},
  "equal x, y":     {x: 4, y: 4, want: "1.00"},
  "max x, pos y":   {x: 127, y: 2, want: "63.50"},
  "min x, pos y":   {x: -128, y: 2, want: "-64.00"},
  "zero x, pos y":  {x: 0, y: 2, want: "0.00"},
  "pos x, zero y":  {x: 10, y: 0, wantErr:
    errors.New("cannot divide by 0")},
  "zero x, zero y": {x: 0, y: 0, wantErr:
    errors.New("cannot divide by 0")},
  "max x, max y":   {x: 127, y: 127, want: "1.00"},
  "min x, min y":   {x: -128, y: -128, want: "1.00"},
}
```

In practice, we would expand each of the edge and corner cases one at a time, ensuring that each of them passes. However, we will add them all in one step, for the purpose of brevity.

Step 9 – expanding functional code

As expected, the new error edge cases will fail when run with the typical `go test` command, prompting us to implement functional code. We expand our `Divide` function to handle the error case described in the user requirements:

```
func Divide(x, y int8) (*string, error) {
  if y == 0 {
    return nil, errors.New("cannot divide by 0")
  }
  r := float64(x) / float64(y)
```

```
    result := fmt.Sprintf("%.2f", r)
    return &result, nil
}
```

As usual, the error case is handled at the top of the function, keeping the code minimally indented. Note that we initialize an error using the `errors.New` function, which takes in a message. We can initialize errors in other ways as well.

The final step is to run our fully implemented table-driven test suite using the `go test` command:

```
$ go test -run TestDivide ./chapter04/table -v
--- PASS: TestDivide (0.00s)
   --- PASS: TestDivide/zero_x,_pos_y (0.00s)
   --- PASS: TestDivide/max_x,_max_y (0.00s)
   --- PASS: TestDivide/max_x,_pos_y (0.00s)
   --- PASS: TestDivide/min_x,_pos_y (0.00s)
   --- PASS: TestDivide/equal_x,_y (0.00s)
   --- PASS: TestDivide/min_x,_min_y (0.00s)
   --- PASS: TestDivide/pos_x,_zero_y (0.00s)
   --- PASS: TestDivide/zero_x,_zero_y (0.00s)
   --- PASS: TestDivide/pos_x,_pos_y (0.00s)
   --- PASS: TestDivide/neg_x,_neg_y (0.00s)
PASS ok        github.com/PacktPublishing/Test-Driven-
Development-in-Go/chapter04/table          0.298s
```

As we can see from the output, all the tests are run successfully in their own subtest, with the given scenario name passed to the test runner. Our first table-driven test suite has been successfully implemented. This is a common testing technique that you will often use when you write Go code, so it's important to master its methods.

Parallelization

By default, all the tests in each package will be run sequentially, but tests from multiple packages will run in parallel. As the number of tests increases, the sequential test execution time of a given package can increase as well.

Figure 4.7 demonstrates the behavior of sequential and parallel test runs:

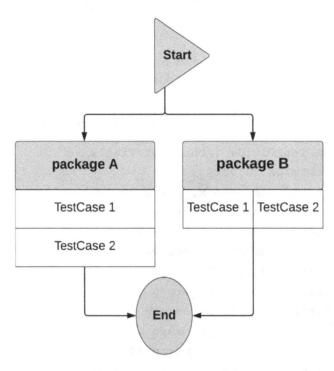

Figure 4.7 – Sequential versus parallel test runs

The test run life cycle is set out here:

- The tests begin running. Tests in different packages run in parallel—tests in `package A` can run at the same time as tests in `package B`. This allows us to reduce running time.

- By default, test cases in the same package run sequentially. This is demonstrated by the tests in `package A`—`TestCase 1` needs to complete before `TestCase 2` can run.

- The test cases in the same package can be configured to run in parallel. This is demonstrated by the tests in `package B`—`TestCase 1` can run concurrently with `TestCase 2`.

The number of tests that we can run in parallel is limited by the resources available to the test runner, but parallelizing test runs is a great way to reduce the test run time, which can further reduce the feedback cycle of CI/CD pipelines.

The `*testing.T` type provides the `t.Parallel()` method, which allows us to specify which tests can be run in parallel with other parallel marked tests from the same package. As the subtests of our table-driven test run independently, we need to mark each as parallel and not just the top-level test.

The ability to mark certain tests for parallelization is particularly useful together with table-driven tests, which contain independently running test cases. We can easily adjust our table-driven tests to run in parallel with two short lines of code:

```
for name, rtc := range tests {
  tc := rtc
  t.Run(name, func(t *testing.T) {
    t.Parallel()
    x, y := int8(tc.x), int8(tc.y)
    r, err := table.Divide(x, y)
    if tc.wantErr != nil {
      assert.Equal(t, tc.wantErr, err)
      return
    }
    assert.Nil(t, err)
    assert.Equal(t, tc.want, *r)
  })
}
```

We assign the current test case to a local `tc` variable to capture the test case range variable. This is required as the subtest will now run in a goroutine under the hood. We need to create a copy of the current value of the test case to the subtest closure, as opposed to the changing range return value.

The second change we have made is to add the call to `t.Parallel()` in the subtest, marking each of the subtests as allowed to be run in parallel.

By default, the number of binaries that can run in parallel is equal to the number of CPUs. This variable can be overridden by the `-parallel` flag, available on the `go test` command.

With our table-driven tests marked as parallel, we can run our tests again using `go test`:

```
$ go test -run TestDivide ./chapter04/table -v
=== RUN    TestDivide
=== RUN    TestDivide/pos_x,_pos_y
=== PAUSE  TestDivide/pos_x,_pos_y
=== RUN    TestDivide/neg_x,_neg_y
=== PAUSE  TestDivide/neg_x,_neg_y
=== CONT   TestDivide/pos_x,_pos_y
=== CONT   TestDivide/neg_x,_neg_y
--- PASS:  TestDivide (0.00s)
```

```
--- PASS: TestDivide/pos_x,_pos_y (0.00s)
--- PASS: TestDivide/neg_x,_neg_y (0.00s)
ok      github.com/PacktPublishing/Test-Driven-Development-
in-Go/chapter04/table          0.223s
```

The output of the test run has been shortened. As we can see from the interleaving output, the tests are now running in parallel, in an interleaving manner: run, pause, and continue.

Advantages and disadvantages of table-driven testing

This brings us to the end of our exploration of table-driven testing. Let us conclude with a short discussion of its advantages and disadvantages. Table-driven tests are best suited to scenarios that cover a variety of test cases with different inputs and outputs.

Advantages

Table-driven tests have the following advantages:

- Provide a concise way to define and run multiple test cases, which reduces boilerplate code
- Easy to add and remove new tests by simply modifying the collection of test cases
- As all of the test cases are run using the same surrounding code, we can easily refactor the test setup and assertion code

Disadvantages

Here are some disadvantages of table-driven tests:

- As all the test cases are run identically, it may be difficult to create even small variations of the test setup and assertion code.
- Table-driven tests are not suitable for test cases that require different test setup and teardown logic. They also make it difficult to use mocks, which must behave differently.
- Some developers argue that table-driven tests are difficult to read. While the name of the test case allows us to name each test, the code is not readable, especially when compared to the **behavior-driven development** (BDD) style of writing tests.

When implemented correctly, table-driven tests are a great way to test your code across a variety of scenarios and edge cases. They help us create a uniform way of running tests, which also makes it easy to maintain and refactor test code. Many developers advocate implementing your tests as table-driven tests from the very beginning, even if you don't have many test cases when you get started. As your code matures, you will have an easy way to add new test cases.

If you have large variations in test setup, you can use different tests and dedicated subtests to group your tests.

Use case – the BookSwap application

One of the most popular use cases of Go is for building web applications. Therefore, it is important to know how to build and test web applications. We will learn how to build our first use case web application: the BookSwap application. We will explore and test the BookSwap application in this and the following chapters.

This simple application allows users to sign up and register which books they have available. Other users can sign up for the application and view other users' available books. They can then request to borrow a book from another user. The BookSwap application then generates an order and sends it to the posting service for wrapping and shipping.

Figure 4.8 depicts an overview of the BookSwap application:

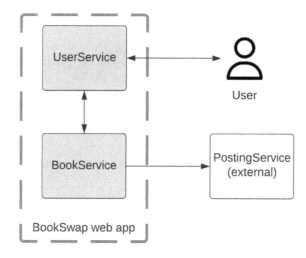

Figure 4.8 – Overview of the book swap web application

The BookSwap web application has some simple components:

- The user interacts with the UserService service endpoints. It exposes a few simple endpoints that provide all the functionality the application requires:

    - GET / returns a welcome message and all the book listings available in the application. This endpoint will serve as the application home page, which will show all the available books that users can swap. For completion, the GET /books endpoint will also return a list of available books.

    - POST /users creates a new user. The user will receive a unique userID value, which they will need to remember for further interactions. For the sake of simplicity, we will not handle user authentication or any security concerns.

- GET /users/{id} returns the book listings of the given user. This endpoint will serve as the profile page of a given user.

- The UserService service relies on the BookService. It manages the details and statuses of all the books available in the BookSwap web application. It exposes the following endpoints:

    - POST /books creates a new book listing on the BookSwap service. This request will take the details of the book to create a JSON request body.

    - POST /books/{id}?user={userId} creates a new request for a particular book and given user. This will create a request to send the given book to a new user.

- BookService has a dependency on the external posting service, which handles the creation of mail stamps and issues a request for packaging. Once PostingService has handled the order request, we can mark the book as swapped and update its ownerID value.

You can explore the full implementation of the BookSwap application on our GitHub repository. The application is implemented using the net/http package in the standard library. We will explore some relevant parts of the BookSwap web application in this chapter, which showcase what we have learned in this chapter.

Testing BookService

We will represent books as a very simple data type that uses JSON tags to format its contents for display on REST APIs, which serve JSON data:

```go
type Book struct {
    ID       string `json:"id"`
    Name     string `json:"name"`
    Author   string `json:"author"`
    OwnerID  string `json:"owner_id"`
    Status   string `json:"status"`
}
```

While REST APIs don't have to operate on JSON data, the application/json data format is the most used. The book has OwnerID and Status, which shows whether the book is available for swaps.

BookService is a very simple service that manages books. It will need to be able to retrieve and manage books with six simple methods:

```go
// NewBookService initializes a BookService given its
// dependencies.
func NewBookService(initial []Book, ps PostingService)
```

```
*BookService

// Get returns a given book or error if none exists.
func (bs *BookService) Get(id string) (*Book, error)

// Upsert creates or updates a book.
func (bs *BookService) Upsert(b Book) Book

// List returns the list of available books.
func (bs *BookService) List() []Book

// ListByUser returns the list of books for a given user.
func (bs *BookService) ListByUser(userID string) []Book

// SwapBook checks whether a book is available and, if
// possible, marks it as swapped.
func (bs *BookService) SwapBook(bookID, userID string) (*Book,
error)
```

The NewBookService method initializes BookService with a given list of books and a PostingService service. The Get method attempts to find a book for a given ID, returning an error if not found. The Upsert method creates a new book entry or updates the entry if the given ID is already found. The List operation returns all the books that are available for loan. ListByUser filters all books for a given owner, allowing us to power the home page of a given user. SwapBook is a function that wraps around availability, checking and updating the owner ID of a given book in case of a swap request.

BookService will save the book entries on a map with their ID as the key:

```
type BookService struct {
        books map[string]Book
        ps    PostingService
}
```

The map will facilitate lookup operations, which will be required for the BookSwap application. The Get and List operations are expected to be the most popular, as they will appear on the homepage and profile pages.

Let us have a look at how we can formulate the table-driven tests for the `Get` operation of the `BookService`. We declare a test with two corresponding subtests—one for an initial amount of books and one for an empty book map:

```
func TestGetBook(t *testing.T) {
  t.Run("initial books", func(t *testing.T) {
   // Books are available in the BookService
  })
  t.Run("empty books", func(t *testing.T) {
    // No books in the BookService
  })
}
```

We use two different subtests for the two cases, as they require two different test setups. As we have discussed, table-driven tests are not suitable for scenarios that require different setup conditions. We begin by creating a sample book and creating a new `BookService` instance:

```
eb := db.Book{
  ID: uuid.New().String(),
  Name: "Existing book",
  Status: db.Available.String(),
}
bs := db.NewBookService([]db.Book{eb}, nil)
```

This starting point will be shared by all test cases in this subtest. Note that we pass a `nil` value as the `PostingService` service, as it will not be tested by these tests. Then, we implement a table-driven test with three scenarios in the first subtest:

```
tests := map[string]struct {
  id string
  want db.Book
  wantErr error
}{
  "existing book": {id: eb.ID, want: eb},
  "no book found": {id: "not-found", wantErr:
    errors.New("no book found")},
  "empty id": {id: "", wantErr: errors.New("no book
    found")},
}
```

The three cases consist of finding an existing book, looking for a book that is not found in `BookService`, and looking for an empty ID. Then, we loop through the test cases and run the assertions according to the inputs and expectations of the test case:

```
for name, tc := range tests {
  t.Run(name, func(t *testing.T) {
    b, err := bs.Get(tc.id)
    if tc.wantErr != nil {
      assert.Equal(t, tc.wantErr, err)
      assert.Nil(t, b)
      return
    }
    assert.Nil(t, err)
    assert.Equal(t, tc.want, *b)
  })
}
```

Just as we did in the *Table-driven testing in action* section, we loop through the map of test cases and handle the error cases first. Remember to verify all the wanted return values.

In the second subtest, named `"empty books"`, we have run a single test and performed the required verifications on a different UUT instance:

```
t.Run("empty books", func(t *testing.T) {
    bs := db.NewBookService([]db.Book{})
    b, err := bs.Get("id")
    assert.Equal(t, errors.New("no book found"), err)
    assert.Nil(t, b)
})
```

We could have potentially implemented table-driven tests for the second subtest as well, but we have opted to include a single test here to keep the code snippets concise.

Finally, we run our tests using the `go test` command to ensure that they are passing:

```
$ go test -run TestGetBook ./chapter04/db -v
=== RUN   TestGetBook
--- PASS: TestGetBook (0.00s)
  --- PASS: TestGetBook/initial_books (0.00s)
  --- PASS: TestGetBook/initial_books/existing_book (0.00s)
```

```
 --- PASS: TestGetBook/initial_books/no_book_found (0.00s)
 --- PASS: TestGetBook/initial_books/empty_id (0.00s)
 --- PASS: TestGetBook/empty_books (0.00s)
PASS
ok       github.com/PacktPublishing/Test-Driven-Development-
in-Go/chapter04/db    0.217s
```

Note that the output shows the nesting of the two different subtests. This allows us to build detailed test hierarchies. We will continue to explore and test other parts of the BookSwap application in the next few chapters, so there will be plenty of time to explore it.

Summary

In this chapter, we explored how to identify edge cases and write test suites that cover multiple conditions. We began with how to identify edge cases for systems with input parameters and external services, revising Go's approach to explicit error handling. Then, we learned how to implement table-driven testing. This popular technique allows us to test multiple scenarios with a minimal amount of boilerplate code. It also allows running test cases in parallel, enabling us to make optimizations for the running of test cases as well. Finally, we introduced our new use case—the BookSwap web application. This example application will be the focus of the next few chapters, where we will learn how to test one of Go's most popular use cases: building web applications.

In *Chapter 5, Performing Integration Testing*, we will begin to consider how to use TDD for testing **end-to-end** (**E2E**) applications, including database testing. We will also learn how to use Docker for identical application setup and easy teardown.

In *Chapter 10, Testing Edge Cases*, we will explore other testing techniques, such as fuzz testing and property-driven testing, which can make edge case verification even easier.

Questions

1. What is an edge case? What is a corner case?

2. What is an idempotent operation?

3. Explain Go's explicit error handling.

4. What is table-driven testing? What are some of its advantages?

5. How does Go parallelize test runs?

Further reading

- *Building Microservices Second edition: Designing Fine-Grained Systems, Sam Newman*, published by *O'Reilly*.

- *Error handling and Go on the Go blog.* Available at `https://go.dev/blog/error-handling-and-go`.

Part 2:
Integration and End-to-End
Testing with TDD

With the fundamentals of TDD and Go testing in place, this part moves our focus beyond testing components in isolation. We explore the importance of integration testing and learn how to write tests using `httptest` and `ginkgo` to test the `BookSwap` web application. Then, we extend the functionality of the application by adding a database and containerizing it using Docker, which allows us to create identical test setups. Once the application is extended into a monolithic application, we use `GoDog` to implement end-to-end testing using BDD-style features. However, refactoring is an integral part of the development process, often applied to splitting monolithic applications to microservice architectures. We examine testing microservice integrations using contract testing, implemented using the Pact open-source testing tool.

This part has the following chapters:

- *Chapter 5, Performing Integration Testing*
- *Chapter 6, End-to-End Testing the BookSwap Web Application*
- *Chapter 7, Refactoring in Go*
- *Chapter 8, Testing Microservice Architectures*

5

Performing Integration Testing

In the previous chapters, we discussed the broader topic of writing and testing code with **test-driven development** (**TDD**), but have kept our implementation focus on unit tests. As we've discussed at length so far, unit tests are at the bottom of the test pyramid, being the most numerous, as they are testing all the different independent parts or units of the application.

The concepts we have discussed so far have allowed us to write unit tests that test these units in isolation, across a variety of scenarios. In *Chapter 3, Mocking and Assertion Frameworks*, we learned how to make use of frameworks to easily create mocks, which allow us to instantiate units whose dependencies we have full control over. As discussed in *Chapter 4, Building Efficient Test Suites*, we learned how to make use of the popular technique of table-driven testing to easily write tests across a variety of cases, including edge and corner cases.

No matter how well we write our unit tests, they have the limitation that they only verify their limited scope. In other words, unit tests verify that each unit is working correctly, but not that they integrate and function correctly together. The integrations between units, which may be developed by different teams, can often be the cause of errors and outages, so it is important to verify that they behave as expected, independently and together.

We will now turn our attention to implementing integration testing suites, which will give us the confidence that the functionality that matters will work as intended when multiple units work together. We will explore how to containerize our applications, ensuring that our tests mimic our production environments and perform as expected.

In this chapter, we will cover the following topics:

- The limitations of unit testing
- The implementation of integration tests in Go
- Introduction to behavior-driven test writing
- The importance of database testing
- Containerization with Docker

Technical requirements

You will need to have **Go version 1.19** or later installed to run the code samples in this chapter. The installation process is described in the official Go documentation at `https://go.dev/doc/install`.

The code examples included in this book are publicly available at `https://github.com/PacktPublishing/Test-Driven-Development-in-Go/chapter05`.

Supplementing unit tests with integration tests

Unit tests are small, fast tests that verify the behavior of a single component. In Go, the UUT is typically the package, which exposes an API that these fast tests can verify against. These independent units combine to make up **components**, which are identifiable parts of a system. Usually, components have well-defined responsibilities and provide a group of related functions. A component's units work together to deliver the component's functionality.

Engineers rely heavily on unit tests in the development phase, and they are an important pillar of TDD, where the testing practice requires the testing code to be written together with the implementation code. However, they have some limitations that make the remaining tests of the testing pyramid essential. Therefore, as TDD practitioners, we cannot simply focus on unit tests.

Limitations of unit testing

The practice of verifying functionality with unit tests has been the subject of debate in the engineering community because of its limitations. *Figure 5.1* presents a summary of their advantages and disadvantages:

Advantages	Disadvantages
+ Support refactoring	- Increase the amount of code
+ Early bug detection	- Increased refactor effort
+ Easier debugging	- Difficult to identify realistic scenarios
+ Better code design	- Difficulties testing user interfaces (UI)
+ Documentation alongside implementation	

Figure 5.1 – Advantages and disadvantages of unit testing

Here are the advantages of unit tests:

- **Support refactoring**: Unit tests make it easier to refactor code because they provide fast verification of existing functionality. They decrease the risk associated with changing code, which can lead to breaking existing functionality.

- **Early bug detection**: Unit tests verify the implementation at the development phase before it has been integrated with the existing product and can be tested end-to-end. This also ensures that bugs don't propagate to other teams or are accidentally released. Early bug detection can also lead to shorter development times and reduced project costs.

- **Easier debugging**: Detecting and fixing errors is easier when the tests have a limited scope. As the **UUT** is tested in isolation from its dependencies, we know that any failing tests are caused either by the test setup or the implementation of the UUT.

- **Better code design**: Poorly designed code is hard to test code and can highlight to developers where their code must be rewritten or refactored. In practice, unit tests promote better code design because they bring the testing concerns to the development phase.

- **Documentation alongside implementation**: Unit tests serve as detailed documentation for the functionality and behavior of a component. As tests live alongside the code in Go, they give developers access to it without the use of another documentation system.

And these are the disadvantages:

- **Increase the amount of code**: Unit tests increase the code that developers must write early on. This is problematic for tasks that require prototyping or don't have well-established requirements. Developers don't want to write large amounts of code that then need to be changed alongside the implementation.

- **Increased refactor effort**: While unit tests ensure that refactoring has not broken any existing functionality, causing regressions, the tests themselves must be refactored in the case of changes in requirements. This can increase the cost of refactoring efforts.

- **Difficult to identify realistic scenarios**: As the codebase grows and functionality becomes more complex, it will be difficult, if not impossible, to test all the execution paths of a component. However, as unit tests are written based on code and not user requirements, it can be difficult for developers to identify which scenarios are realistic and should be covered.

- **Difficulties testing user interfaces (UIs)**: It is difficult to test UIs with unit tests. Usually, they verify business logic, as they traditionally do not have libraries available for UI verification.

Integration tests are a good way to supplement unit tests, as they address some of the disadvantages and limitations of unit tests highlighted previously. Next, we will learn how to implement and run them for our Go packages.

> **Unit tests are considered good practice**
>
> While they do pose some disadvantages, the consensus in the community is that they should be used as part of development practice. Understanding their limitations highlights what other testing needs we need to cover for the full verification of our system.

Implementing integration tests

Integration tests and end-to-end tests are often used interchangeably, but they each have a scope and purpose in the testing pyramid. *Figure 5.2* depicts the testing pyramid and highlights the difference in scope and speed between integration and end-to-end tests:

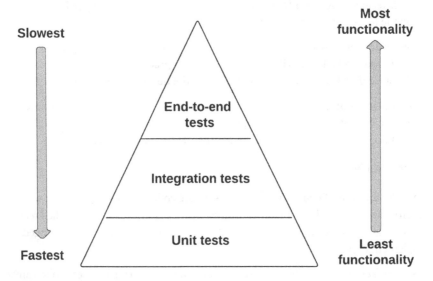

Figure 5.2 – The distinction between integration and end-to-end tests

The difference in speed between integration and end-to-end tests is due to the functionality that they cover:

- **Integration tests** cover one or multiple components, ensuring that the individual components work well as a combined entity. While the logic of the particular component is verified by its unit tests, the purpose of the integration test is to exercise the conditions at the seams between the components.

- **End-to-end tests** replicate the usage of the system by the user. They require starting up all services and dependencies of the system under test. Then, tests that mimic user behavior are written using helper frameworks. These tests verify that the system is performing correctly under real-world conditions.

So, if end-to-end tests cover more functionality than integration tests and can be automated, why should we bother to implement integration tests? *Figure 5.3* depicts some of the drawbacks of end-to-end tests and how integration tests address them:

Integration tests	End-to-end tests
+ Performed early in the development process, shortening feedback cycle + Faster and cheap to run + Test integration with external and internal modules	- Performed at the end of the development process, when the product is complete - Slow and possibly expensive - Test user flow and experience

Figure 5.3 – Challenges of end-to-end tests

All of the tests in the testing pyramid work together to address each other's shortcomings. In particular, integration tests and end-to-end tests work together in unison:

- Typically, end-to-end tests are performed at **the end of the development process**, once the system is relatively stable and can be called end to end. On the other hand, integration tests can be performed as soon as the individual components are ready, earlier in the development cycle, thereby **shortening the feedback loop** and allowing developers to detect bugs earlier on in the project.

- As they require more setup and resources, end-to-end tests are **slow and possibly expensive** to run. Therefore, engineers might run them as releases and not individual code commits. On the other hand, integration tests require much less setup, so they are **faster and cheaper to run**. They are often included in the code commit checks.

- As previously mentioned, the focus of end-to-end tests is to verify the **test user flow and experience** in real-world scenarios. On the other hand, integration tests focus on **integration with external and internal modules** in a variety of scenarios, such as negative testing and partial outages. These can be difficult to set up in end-to-end tests, which require the entire system to be configured.

> **Integration tests are implemented just as unit tests**
>
> We use the same mechanisms for integration tests. We make use of setup functions and mocks and table tests to write tests that simply have a larger scope. Furthermore, integration tests have the same test signature as unit tests.

The setup for integration tests is slightly more complex than unit tests, as multiple components, some of which are external, must be configured and started. *Figure 5.4* shows a typical example of the technologies and configurations we might use:

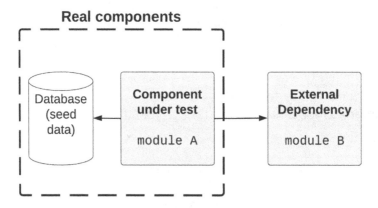

Figure 5.4 – Example configuration of integration tests

The various parts of the integration tests that need to be configured are as follows:

- The **Component under test** part is initialized. The component under test is larger than the UUT, but it is still self-contained and defined within a single module. The scope of the integration test is to ensure multiple units work as expected, but they are always contained within the single module under test.

- If required, we initialize the **Database** component with a given seed/start position of test data contained inside it. As they are complex, databases are rarely mocked and will most often be started and populated before the component under test is started. Database start positions are often specified as **SQL files** or **JSON files**.

- Docker makes it easier to configure **Real components** together and is often used for system configuration. We will look at how to leverage the power of Docker later in this chapter in the *Spinning up and tearing down environments with Docker* section.

- Most often, the component under test will require dependencies for it to start and function correctly. These dependencies could be internal to the project or external dependencies to the organization, such as a third-party service. These external dependencies will be mocked, allowing us to test our component with a variety of inputs and conditions.

Let's have a look at an example integration test for our BookSwap application, which we introduced in *Chapter 4*, *Building Efficient Test Suites*. We will write an integration test for the GET / endpoint that will return a welcome message and a list of available books. It will also allow us to explore testing web applications.

The HTTP handler that's registered to respond to this request is relatively simple:

```
// Handler contains the handler and all its dependencies.
type Handler struct {
  bs *db.BookService
  us *db.UserService
}
// Index is invoked by HTTP GET /.
func (h *Handler) Index(w http.ResponseWriter, r *http.Request)
{
  // Send an HTTP status & a hardcoded message
  resp := &Response{
    Message: "Welcome to the BookSwap service!",
    Books:     h.bs.List(),
  }
  writeResponse(w, http.StatusOK, resp)
}
```

The implementation of Handler highlights the following implementation details:

- We create a custom Handler type with all its required dependencies. In the case of the BookSwap application, we save an instance of BookService and an instance of UserService.

- The handler has a method for each endpoint that it serves. We create a handler method that takes in a ResponseWriter and a Request. This signature is typical of http.HandlerFunc, which is an adapter to allow the use of Go functions as HTTP handlers.

- We invoke the List function of BookService to fetch the list of books and construct a response. This custom response is then written to ResponseWriter, which allows us to easily unmarshal Go structs to HTTP responses.

The setup of our handler code is pretty straightforward and will be similar to any code you will write for HTTP responses. But how would we test it? We could unit test BookService and ensure that it functions correctly, but we also need to test that the responses the handlers construct are as expected. It's time to write our very first integration test.

The Go standard library has the `httptest` package (`https://pkg.go.dev/net/http/httptest`), which allows us to easily test HTTP handlers and clients. This package contains the functionality for the following:

- Starting servers with a specific `http.HandlerFunc` with the `httptest.Server` type.

- Creating incoming requests to pass to handlers with the `httptest.NewRequest` function.

- Recording responses with the `httptest.ResponseRecorder` type for assertions in testing code. The recorder conforms to the `http.ResponseWriter` type and can be used in its place in handler code.

A simple integration test for our `GET / ` HTTP handler is as follows:

```go
func TestIndexIntegration(t *testing.T) {
 // Arrange
 book := db.Book{
  ID: uuid.New().String(),
  Name: "My first integration test",
  Status: db.Available.String(),
 }
 bs := db.NewBookService([]db.Book{book}, nil)
 h := handlers.NewHandler(bs, nil)
 svr := httptest.NewServer(http.HandlerFunc(h.Index))
 defer svr.Close()
 // Act
 r, err := http.Get(svr.URL)
 // Assert
 require.Nil(t, err)
 assert.Equal(t, http.StatusOK, r.StatusCode)
 body, err := io.ReadAll(r.Body)
 r.Body.Close()
 require.Nil(t, err)
 var resp handlers.Response
 err = json.Unmarshal(body, &resp)
 require.Nil(t, err)
 assert.Equal(t, 1, len(resp.Books))
 assert.Contains(t, resp.Books, book)
}
```

The `TestIndexIntegration` test is relatively straightforward since it does not require any complex request construction or response verification:

1. The signature of the test is just like any other unit test. It starts with the `Test` prefix and takes in a single parameter of the `*testing.T` type.

2. Next, we create an instance of `BookService` with a single book as the starting position. The purpose of the test is to ensure that `BookService` integrates with its handler and returns responses as expected.

3. We create a new handler with the instantiated `BookService`. Then, we pass the handler to the `httptest.NewServer` function, which creates and starts a server instance to serve our handler. We defer the call to the `Close` function, as this server should be shut down at the end of the test execution. This concludes the `Arrange` section of our test.

4. The `Act` section of our test is very simple. We invoke the server at its URL using the `http.Get` method. This is the same method that our clients will be using, and the test is not aware that it is calling a special, mocked server.

5. Finally, we can run assertions on our response and possible error in the `Assert` section of our test. We verify that no error is returned and that the response has the `200 OK` HTTP status code.

6. Then, we read the body of the response and unmarshal it into our custom response type. This makes it easier for us to verify the response, but we could have also verified the contents of the response body as a string.

7. The last assertion verifies that the book instance created in the `Arrange` section is contained in the custom response. The test then concludes and the deferred call to the server `Close` function is run, cleaning up the server resources set up by the test.

The `httptest` package allows us to seamlessly verify the behavior of HTTP handlers and integration tests using the same libraries and functions that clients will use. This allows us to write powerful integration tests.

Running integration tests

Integration tests can be run just like any other unit test that we have been running so far – by using the `go test` command:

```
$ go test -run TestIndexIntegration ./chapter05/handlers -v
=== RUN    TestIndexIntegration
--- PASS: TestIndexIntegration (1.712s)
PASS
ok      github.com/PacktPublishing/Test-Driven-Development-
in-Go/chapter05/handlers      1.712s
```

The test runs successfully since it has the typical signature of a unit test. However, notice that this integration test takes nearly 2 seconds to run on my machine. This is the measurement for a particular test run, but I have registered runtimes as high as 4 seconds for just this simple GET request. As the number of integration tests for a particular application grows, they have the potential to severely slow down our test suites, even if we run them using t.Parallel(), as we learned in *Chapter 4, Building Efficient Test Suites*.

It would be great to separate our unit tests and much slower integration tests. We could then run unit tests for all commits and integration tests for code releases. There is no perfect, built-in way to signal to the test runner which tests are integration tests, but we can explore a few options.

Short mode

The go test command has a built-in flag called -short that we can access using the testing. Short() function. This flag allows us to mark long-running tests for skipping by adding a short snippet to their test code:

```
func TestIndexIntegration(t *testing.T) {
  if testing.Short() {
   t.Skip("Skipping TestIndexIntegration in short mode.")
  }
  // testing code continues
}
```

The t.Skip method will ensure that this long-running test will be skipped. We can then run the tests in short mode by adding the -short flag to our test command:

```
$ go test -run TestIndexIntegration ./chapter05/handlers -v
-short
=== RUN    TestIndexIntegration
    handlers_test.go:19: Skipping TestIndexIntegration in short
mode.
--- SKIP: TestIndexIntegration (0.00s)
PASS
```

As expected, the long-running test is skipped.

The major downside of this approach is that it requires the user to have special knowledge to achieve a fast-running test suite, which should be the default behavior. There is no built-in -long flag that we can use to execute all (including long-running) tests.

Naming conventions

Another option is to use naming conventions, which would not require any special code functions to be added to any tests. For example, you could agree with in your team that unit tests will end with the Unit suffix and integration tests with the Integration suffix. Depending on the length and contents of the file, we could create separate integration and unit test files. Both unit and integration tests can use the dedicated test package, named with the _test suffix, keeping the source and test code dependencies separated.

Then, we can make use of the -run flag, which we explored in *Chapter 2, Unit Testing Essentials*, to instruct the test runner to run a subset of tests based on their name. We run all unit tests using the go test -run Unit ./... command, which will recursively traverse folders to search for any test that contains the word Unit. Analogously, integration tests will be run using the go test -v -run Integration ./... command.

Unfortunately, this method suffers from the same major downside as short mode, as running the default go test command without the -run flag will cause all tests to run, including the slower integration tests.

Environment variables

The last option is to create an environment variable to make up for the lack of a corresponding flag. Again, we will have to add a short code snippet to our test to verify this environment variable:

```
func TestIndexIntegration(t *testing.T) {
  if os.Getenv("LONG") == "" {
    t.Skip("Skipping TestIndexIntegration in short mode.")
  }
  // testing code continues
}
```

We make use of the os.Getenv method to read environment variables, which will return empty if the variable has not been defined. If this variable is empty, we skip the integration test, allowing the default behavior of our test suite to only run fast tests, skipping integration tests.

Running integration tests is easy:

```
$ LONG=true go test -run TestIndexIntegration ./chapter05/
handlers -v
=== RUN    TestIndexIntegration
--- PASS: TestIndexIntegration (0.00s)
PASS
ok      github.com/PacktPublishing/Test-Driven-Development-
in-Go/chapter05/handlers    0.779s
```

Note that this version of the command will only run on CMD terminals. Alternatively, you can set the LONG environment variable to 'true' in your terminal and then run the preceding go test command on its own after setting this.

We will make use of the environment variables solution going forward. The expected default behavior of the test suite is to run fast-running unit tests. This solution allows us to keep specialized knowledge out of the expected default behavior and makes it easy to run integration tests when required. It also integrates well with containerization technologies such as Docker, which we will explore later in this chapter.

Behavior-driven testing

We have now learned how to supplement unit tests with integration tests, increasing the scope of our component under test. End-to-end tests have the most scope as they test the entirety of our system. They are often discussed together with **behavior-driven design** (**BDD**), which is a branch of TDD that focuses on writing human-readable tests based on user requirements.

Fundamentals of BDD

The first step of BDD practitioners is to establish a shared vocabulary between the different interested parties: business stakeholders, domain experts, and various other engineering functions.

Based on this shared and well-understood vocabulary, the user requirements are then converted into **user acceptance tests** (**UATs**). These tests are end-to-end tests that ensure that system requirements are covered by all new releases.

Tests are usually written in the **Given-When-Then** structure, using business language and the shared vocabulary previously established by the business. A BDD formulation of the integration test for the GET / endpoint we have previously implemented looks like this:

- **Story: View the list of books**
- **Given** a user
- **When** the user accesses the GET / root endpoint
- **Then** the list of available books is returned to the user

The test specification reads like plain English and establishes the main aspects of the test case:

- Who the main actor of the test case is
- What their expected behavior is
- What the user will get from the performed action

Note that the test case does not specify any implementation details of the application and instead focuses on *the behavior* of the application. Test cases treat the application as a **black box**. This simplicity is the power of BDD, where test specifications are not something that only engineers and testing professionals can write.

> **BDD is about bridging gaps**
>
> The emphasis on shared language and easily readable tests ensures that the gaps between technical and non-technical stakeholders are bridged. This avoids misunderstandings and delays in the implementation of the system's intended behavior.

Figure 5.5 highlights some of the advantages and disadvantages of writing tests using BDD:

Advantages	Disadvantages
+ Single source of truth + Tests as documentation + Specific behaviors + Wider involvement	- Time consuming - Requires commitment - Dependent on good BDD practices

Figure 5.5 – Advantages and disadvantages of writing BDD tests

The following are the advantages of BDD:

- **Single source of truth**: The biggest advantage of BDD is that it allows teams to have a single source of truth for the intended behavior of the application. Furthermore, we have a unified vocabulary to express this behavior across the business.

- **Tests as documentation**: While unit tests can also serve as documentation for the application, BDD tests are easier to read and understand, since they focus on readability.

- **Specific behaviors**: With their Given-When-Then structure, BDD tests encourage writing test cases for specific behavior. This often helps narrow down larger and potentially vague user requirements that have been established at the beginning of the project.

- **Wider involvement**: Anyone in the team or the business can contribute to the specification of these tests, making it easier to detect any bugs or functional oversights early on.

And here are the disadvantages:

- **Time-consuming**: It can be time-consuming to get multiple stakeholders together to establish test cases at the beginning of the project. Furthermore, it can also be time-consuming to maintain these tests during the lifetime of the project.

- **Requires commitment**: The different stakeholders need to commit to taking on the work of specifying and discussing these test cases upfront.

- **Dependent on good BDD practices**: Unless correctly specified together with the correct stakeholders, BDD tests can become ambiguous and difficult to implement. The successful specification of tests is therefore dependent on good BDD practices in the business.

Now that we understand some of the advantages of BDD tests and how to write them, we can turn our attention to implementing them in Go.

Implementing BDD tests with Ginkgo

In *Chapter 3, Mocking and Assertion Frameworks*, we learned how to create mocks and write assertions with the testify open source testing library. This allows us to create streamlined unit tests and easily create mocks. However, a more expressive testing library was required to easily produce BDD-style tests.

The ginkgo (https://github.com/onsi/ginkgo) project was started in 2013 to fill this need. It is a testing framework built on top of Go's testing package and it is designed to help us write expressive BDD tests. It is used together with the gomega (https://github.com/onsi/gomega) matcher library, which exposes assertion matchers that we can use in our tests. This framework received mixed support from the community, as it brought the Ruby way of writing tests to Go. However, it is currently the default way to write BDD-style tests and it is an important part of our TDD journey.

The Ginkgo library supports Go modules and can easily be installed with the go install command, just like testify:

```
$ go get github.com/onsi/ginkgo/v@v2.4.0
$ go install -mod=mod github.com/onsi/ginkgo/v2/ginkgo@v2.4.0
```

> **Ginkgo installation location**
> The install command will install the ginkgo CLI in your $GOBIN path, so ensure that it is set accordingly before you install it. By default, the $GOBIN path is $GOPATH/bin.

The go get command then fetches the gomega assertion library:

```
$ go get github.com/onsi/gomega/...
```

Ginkgo tests live in _test.go files, just like regular unit tests, but they are organized in test suites. Suites can be compared to the table tests that we previously implemented, where we grouped tests by similar functionality and scenarios.

Suites are generated in the current directory using the ginkgo bootstrap command:

```
$ cd chapter05/handlers && ginkgo bootstrap
Generating ginkgo test suite bootstrap for handlers in:
        handlers_suite_test.go
```

The file is named according to the package declared in the current directory. The generated file contains the package declaration and some essential code for the suite's declaration. Note that this command will fail if a suite already exists.

The bootstrap command is a convenient way to generate this boilerplate for us and ensure that all test files have the same basic structure, across our all projects. It also ensures that our suites' naming is consistent, so it is a powerful standardization tool.

> **Testing terminology**
>
> ginkgo uses the same terminology as the **Ruby** community. A **suite** is a collection of tests that all verify the same package. A test is called a **spec**. We will use the same terminology when referring to ginkgo tests going forward.

The generated chapter05/handlers/handlers_suite_test.go, contains the following code:

```go
package handlers_test

import (
  "testing"
  . "github.com/onsi/ginkgo/v2"
  . "github.com/onsi/gomega"
)


func TestHandlers(t *testing.T) {
  RegisterFailHandler(Fail)
  RunSpecs(t, "Handlers Suite")
}
```

This file contains the necessary information for interacting with the `ginkgo` runner:

1. The suite test file is declared inside the `handlers_test` package corresponding to this directory. The separate `_test` package ensures that we only test the exported functionality of the source package. This is essential to writing integration tests that only assert the external behavior of the API.

2. The `ginkgo` and `gomega` libraries are imported using the dot (`.`) operator. This allows us to have access to test and assertion functionality without having to qualify each function with the package name. This can be disabled, but it is discouraged by the BDD community, as tests should read as naturally as possible.

3. The signature of the test is as expected. The test signature takes in a single parameter of the `*testing.T` type. This is the entry point of our generated suite.

4. The test contains two calls to the `Ginkgo` test runner. We will not spend too much time discussing the internals of these functions, but, as all the testing library is open source, you can look up what they do yourself. The call to `RunSpecs` instructs the test runner to begin running the suite and execute all existing specs.

The suite only serves as an entry point for the specs to begin executing, which are usually defined in separate test files.

We define `ginkgo` equivalent to the `Index` endpoint integration test that we previously saw in the *Implementing integration tests* section in the `chapter05/handlers/handlers_index_test.go`:

```go
var _ = Describe("Handlers integration", func() {
 var svr *httptest.Server
 var book db.Book
 BeforeEach(func() {
  book = db.Book{
   ID: uuid.New().String(),
   Name: "My first integration test",
   Status: db.Available.String(),
  }
  bs := db.NewBookService([]db.Book{book}, nil)
  ha := handlers.NewHandler(bs, nil)
  svr = httptest.NewServer(http.HandlerFunc(ha.Index))
 })
 AfterEach(func() {
  svr.Close()
 })
```

```
Describe("Index endpoint", func() {
  Context("with one existing book", func() {
   It("should return book", func() {
    r, err := http.Get(svr.URL)
    Expect(err).To(BeNil())
    Expect(r.StatusCode).To(Equal(http.StatusOK))
    // … assertions continue
   })
  })
 })
})
```

The Ginkgo equivalent of our Index integration test seems quite different from the code we are used to seeing. Its focus is on setting up the various aspects of the test in an easy-to-read **spec tree**:

- We make use of closures to set up our spec hierarchy. The Describe function allows us to create **container nodes**. Specs must begin with a top-level Describe node.

- The BeforeEach function creates **setup nodes** that run before tests. They are used for extracting common setups, allowing us to streamline our tests.

- The AfterEach function creates setup nodes that run after tests. They allow us to clean up after our specs have run, ensuring that critical resources are cleaned up correctly.

- We can further define container nodes inside the top-level nodes as required to organize our specs and their scenarios.

- The Context function is an alias for Describe that allows us to add extra information to our specs to help people understand them. It also creates container nodes but can be used to organize information.

- The It function allows us to define **subject nodes**. These nodes contain the assertions of the subject under test and cannot contain any other nested nodes.

- The assertions inside the subject nodes are written with the gomega assertion library. These can be nested just like the assertions of testify but take a human-readable form. All assertions must begin with the Expect function, which wraps an actual value.

Figure 5.6 shows a visual representation of the structure of the spec tree:

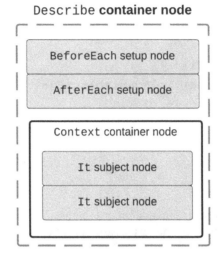

Figure 5.6 – The structure of the spec tree

All tests begin with a `Describe` container node. This top-level node can then contain multiple `BeforeEach` setup nodes, multiple `AfterEach` nodes, other `Context` container nodes, and multiple `It` subject nodes. As we saw in the handlers integration test, these nodes are arranged to build a hierarchy that reflects our test scenario.

> **Nesting rules**
>
> The spec tree consists of nested container nodes. Setup nodes can be nested inside them. Like the behavior of deferred functions, the innermost function will run first. Then, the others will continue in the same fashion going outward.

Once we have generated our suite and populated it with specs, we can run it by using the `ginkgo` command:

```
$ ginkgo -v ./chapter05/handlers
Running Suite: Handlers Suite
================================================================
=====
Handlers integration Index endpoint with one existing book
  should return book
 SUCCESS! -- 1 Passed | 0 Failed | 0 Pending | 0 Skipped
PASS
```

Just like the `go test` command we have used so far, `ginkgo` also supports the `./...` operator, which will traverse subdirectories and look for suites to run.

As we can see from the output, the container nodes and subject nodes are used to construct meaningful names for the spec suite. Ginkgo allows us to construct test collections with meaningful test outputs. We will continue to explore it in future chapters.

Understanding database testing

In the world of testing, databases are often overlooked in literature. Most applications often assume in-memory data storage, just as we have done with the `BookSwap` application so far. However, it is important to understand the difficulties and techniques that we have available when it comes to verifying our databases.

Databases are often seen as external systems or black boxes in our system. They provide specialized behavior and are often complex systems, which most often cannot fail. *Figure 5.6* depicts the typical data translation between different formats:

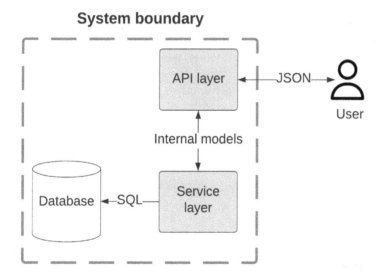

Figure 5.7 – The data formats of a typical system

Data changes formats multiple times in a typical application. User requests usually enter our system in **JSON** format. The **API layer** area then translates these requests to the internal application models and sends them further down the stack to the **Service layer** area. Finally, the **Service layer** area persists these in the database using **SQL** or whatever the expected format of the database is. Often, **NoSQL databases** will save their data back in **JSON** format and persist it.

We should write tests that cover the following aspects of our database systems:

- **Startup and availability**: The application should wait for the database to become available and should do so in an efficient manner.

- **Persistence and querying**: Once data is stored in the database, it should be correctly stored and fetched. This is done by the business logic and should be verified to be implemented correctly.

- **Performance testing**: This type of non-functional testing is important for the database, which typically powers all the requests in the application. Typical verifications include load testing using large files or results counts, running tests using multiple remote users, and any edge cases regarding the values of the column/field values of the database payloads.

These crucial aspects of our systems should be covered by testing, especially around the points where data formats vary and translations occur. These format translations can be the cause of bugs and outages. For example, one field might be a mandatory non-nullable value at the database level, but be missing further up the stack.

> **Mocks as databases**
>
> It might be tempting to assume that a mock would be fitting to wrap around complex external behavior, but the community generally discourages this as an engineering anti-pattern. End-to-end and integration tests should verify and run against the databases that they use in production to avoid differences in functionality and performance.

Useful libraries

Fortunately, the Go ecosystem provides some great libraries to allow us to easily integrate databases into our applications. Here are some Go libraries that you will find useful when integrating databases into your Go applications:

- `go-testfixtures` (https://github.com/go-testfixtures/testfixtures): An open source library that makes it easy to write functional database tests. It uses the **Ruby on Rails** way of setting up data samples using fixtures files.

- `golang-migrate` (https://github.com/golang-migrate/migrate): An open source library that makes setting up database startup positions easy, without us having to write our own data formats and files. It supports a variety of SQL and NoSQL databases.

- `go-txdb` (https://github.com/DATA-DOG/go-txdb): An open source library that runs database queries in transactions. Once the tests are complete, transactions are rolled back and data is not persisted. This allows us to run our tests in isolation on top of a real database.

- `gorm` (`https://github.com/go-gorm/gorm`): A popular open source library that provides **object-relational mapping** (**ORM**). This developer-friendly library makes it easier to convert database types into useful custom structs.

- `bun` (`https://github.com/uptrace/bun`): This is the new, rewritten version of the `go-pg` (`https://github.com/go-pg/pg`) project. This project provides ORM functionality for multiple SQL databases.

The literature on whether to use SQL or NoSQL databases is vast and involves a wide set of recommendations. We will not start this discussion here, but SQL databases remain the most popular database solutions. We will focus on how to implement and test SQL databases in our discussions going forward. We will also see some of the libraries we mentioned in this section in action going forward.

Spinning up and tearing down environments with Docker

The final topic we will cover in our exploration of integration and end-to-end testing is **containerization** using the popular technology known as Docker. It provides us with the ability to start up applications in our local and remote environments in the same way.

Docker gives developers the confidence that their applications will behave in the same way across environments, which is particularly useful for managing and deploying test environments.

Fundamentals of Docker

A **container** is a unit of software that bundles up code and all its dependencies, enabling us to run it in multiple environments. The specification of the container is known as a **container image**. Docker Engine interprets the specification of container images and turns them into containers.

> **Containerization versus virtualization**
>
> **Virtualization** refers to running multiple operating systems on a single machine. Containerization refers to running multiple applications developed in the environment of one operating system on a single machine.

Figure 5.7 depicts containers running on the host environment:

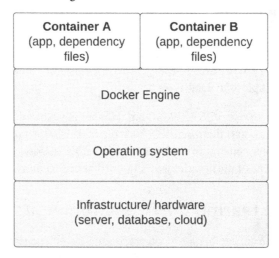

Figure 5.8 – Running containers with Docker

Containers are lightweight, allowing multiple containers to run on the same physical hardware. **Docker Engine** oversees managing them and enforcing isolation levels, ensuring that malicious code cannot escape outside of its current namespace, but also importantly ensuring that tests have a realistic level of isolation. In practice, this means that we can download and run a set of images for complex systems to run on shared hardware.

For example, we could run multiple containers on shared infrastructure: a Go web application, a database, event buses and queues, monitoring, and so on. All these different technologies and images can be managed by a single **standardized technology** with Docker Engine.

Using Docker

The concept of containerization does not belong to Docker exclusively, but we will refer only to the usage of Docker going forward. Docker Engine can easily be installed by following the official documentation available at `https://www.docker.com/get-started/`.

Docker Engine ships with a powerful CLI that contains two main commands:

- `docker` deploys and manages a single application or container. The Docker CLI offers an extensive list of commands and options, some of the most common being these:

  - `docker pull` downloads an image from the image repository, named **Docker Hub** (`https://hub.docker.com/`). Once downloaded, the image is available for use by containers locally.

- `docker run` creates a container from an image. If the image is not available locally, it will be downloaded from the image repository, prolonging the container startup time.

- `docker ps` lists all the locally running containers. This command is commonly used to get the unique container ID for each container. These unique IDs can then be used to reference specific containers in other commands.

- `docker stop` instructs the container to shut down, giving it time to gracefully shut down and clean up its resources. Docker Engine makes use of operating system signals to communicate shutdown to containers. Containers can then be restarted using the `docker start` command.

- `docker kill` instructs the container to stop its execution immediately, without allowing time for graceful shutdown.

- `docker exec` allows us to access a running container. Since containers are isolated from the rest of the operating system, the only way to have access to its resources and setup is to request access from Docker Engine.

- `docker compose` deploys and manages multiple containers within the same single host. This allows us to configure and start multiple containers with a single command, as opposed to starting them each individually with the `docker` command. Another key advantage is that the containers will be running and networking as a single group, making it easy to deploy complex systems across environments. Some of the most common `docker compose` commands are as follows:

  - `docker compose up` starts the specified containers of a given `.yml` file.

  - `docker compose ps` lists the containers of a Compose project, including their statuses and registered ports. These containers will also be visible when running the `docker ps` command, but this command will output more container information.

  - `docker compose stop` instructs running containers to stop, without removing them. They can then be restarted again using the `docker compose start` command.

  - `docker compose kill` forces containers to immediately stop using the `SIGKILL` system signal.

That's all the basics we need to know to install Docker Engine and perform some basic tasks with Docker. In *Chapter 6, End-To-End Testing the BookSwap Web Application*, we will look at the configuration of the custom Dockerfile for our application, the changes we need to make to our existing implementation to make use of a database, and how to run end-to-end tests based on these easy to spin up and tear down containers.

Summary

In this chapter, we moved on from focusing on unit tests, which verify the functionality of code in small isolated units. We began with an introduction to the importance of integration testing and learned how to write and run integration tests for HTTP handlers using the `httptest` library. Then, we learned what the practice of writing BDD-style tests entails and how to implement them using the `ginkgo` testing library. Then, we discussed the importance of testing databases and what useful libraries there are available to us to be able to write these. Finally, we covered the advantages of containerization and learned how to use Docker and configure services with `docker compose`.

In *Chapter 6, End-To-End Testing the BookSwap Web Application*, we will expand on all the fundamentals of the technologies we have learned so far and apply them to test the `BookSwap` web application. This will give us good hands-on practice to configure a typical web application that has a simple database dependency.

Questions

Answer the following questions to test your knowledge of this chapter:

1. What is the difference between integration tests and end-to-end tests?
2. What is **behavior-driven design (BDD)**?
3. Should we mock databases? Why/why not?
4. What is a container?

Further reading

To learn more about the topics that were covered in this chapter, take a look at the following resources:

- *BDD in Action: Behavior-driven development for the whole software lifecycle*, by John Ferguson Smart, published by Manning Publications
- *Docker: Up & Running: Shipping Reliable Containers in Production*, by Sean Kane and Karl Matthias, published by O'Reilly
- *Designing Data-Intensive Applications: The Big Ideas Behind Reliable, Scalable, and Maintainable Systems*, by Martin Kleppmann, published by O'Reilly

6

End-to-End Testing the BookSwap Web Application

We have made a lot of progress toward our goal of learning how to use TDD for implementing and testing Go code. So far, we have covered a wide variety of techniques for implementing unit and integration tests. Unit tests verify that each component works as intended, while integration tests extend their scope to cover the seams and interactions between different units. Moreover, we have learned how to apply these techniques to a wide variety of examples, including the BookSwap web application introduced in previous chapters.

These tests give us a fast feedback loop for code changes, as they don't require the entirety of the application to be started up and made available before the test suite is run. As discussed in *Chapter 5, Performing Integration Testing*, we learned how to make use of the httptest and ginkgo libraries to easily write and run integration tests for web applications. We also learned how to make use of **behavior-driven development** (**BDD**) for writing tests, which is a popular technique for writing integration and **end-to-end** (**E2E**) tests.

However, while we can rely on unit and integration testing to ensure that the application is functioning correctly in a wide variety of scenarios, we should not neglect the benefits that E2E tests bring to our testing strategy. Only E2E tests allow us to verify the behavior of the entire application and replicate the user flows and experience. Simply put, these tests give us an insight into the user experience in production, which is the final verification that we should be making to our application before it is released.

This chapter is dedicated to discussing the implementation of E2E testing suites for the `BookSwap` web application introduced in previous chapters. We will make use of Docker to streamline the creation and teardown of identical applications, as well as discuss the changes we need to make to the application in order to make use of a database. Then, we will learn how to make use of Godog to write and run E2E tests. Finally, we will discuss which database assertions we should include in our tests.

In this chapter, we will cover the following topics:

- The requirements of the `BookSwap` application
- The implementation of database storage in web applications
- Getting started with Cucumber and Godog
- The implementation of E2E tests
- Database start positions and assertions

Technical requirements

You will need to have **Go version 1.19** or later installed to run the code samples in this chapter. The installation process is described in the official Go documentation at `https://go.dev/doc/install`.

The code examples included in this book are publicly available at `https://github.com/PacktPublishing/Test-Driven-Development-in-Go/chapter06`.

Use case – extending the BookSwap application

The `BookSwap` web application was introduced in *Chapter 4, Building Efficient Test Suites*. Its main functionality allows users to list their books and swap them with other users. We learned what its main components and endpoints are, as well as how to apply table testing to its `BookService`. Then, in *Chapter 5, Performing Integration Testing*, we learned how to write integration tests for its `Index` request handler. We will continue to build out the functionality of this application in this chapter, taking a closer look at the user flows and intended functionality for each endpoint.

Figure 6.1 depicts a summary of the responsibilities of the three main services of the `BookSwap` application—`BookService`, `UserService`, and `PostingService`:

BookService	UserService	PostingService
• Creating books • Updating books • Listing books • Filtering by user ID	• Creating users • Updating users • Fetching user by ID and their owned books	• External service • Handling all the posting functionality

Figure 6.1 – The responsibilities of the main components of the BookSwap application

Each of the services has its own specialization and separate responsibilities:

- `BookService` is in charge of all book management aspects. This service implements the functionality of creating, updating, listing, and filtering books. As this application is quite limited in functionality, books are only filtered by the owner's user ID, and we will not implement any searching for the `books` inventory.

- `UserService` is in charge of all user management aspects. This service implements functionality for creating and updating user profiles. It can also fetch a given user by ID and relies on `BookService` to receive a list of all books whose owner ID corresponds to the supplied user ID.

- `PostingService` is an external service to the `BookSwap` application, which is in charge of the details of posting and swapping books. For implementation purposes, we will use a **stubbed implementation** of this service inside the `BookSwap` application. `PostingService` is not a service that actually exists, but we will use an internal stub to simulate calling out to an external service.

> **What is a stub?**
>
> A **stub** is a concrete implementation of another component. Stubs don't make use of mocking frameworks, as they are used by implementation code. They make testing easier and allow us to build code as if the external component were built and implemented. Due to the flexibility of interfaces in Go, stubbed implementations can be easily swapped out for real implementations.

As mentioned in *Chapter 4, Building Efficient Test Suites*, BookSwap saves its data in maps and does not currently have any database or persistent storage. We will change its implementation to use a **PostgreSQL** database in this chapter.

User journeys

In this chapter, we will focus on the implementation of E2E tests. The focus of these tests is to verify the behavior of the application under typical **user journeys**. Therefore, it is important to establish typical user journeys or request flows before writing any E2E test cases.

> **What is a user journey?**
>
> A user journey is the path or sequence of requests that the user of an application will take in order to achieve their goal. Often, these journeys are tracked in production environments to get insights into how users are using services.

Figure 6.2 depicts the expected request flow for a new user for the BookSwap application:

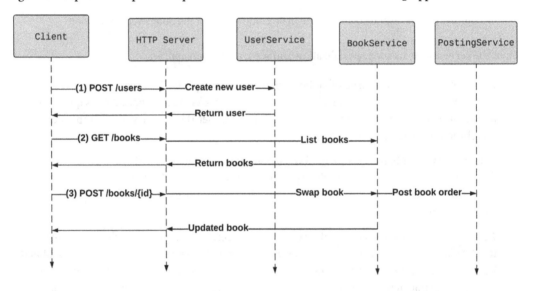

Figure 6.2 – The request flow of a new BookSwap application user

The request flow diagram gives us an insight into which parts of the application are required for the user journey to be successfully completed. The expected usage of the application for a new user is outlined here:

1. **Create a new user**: The user will need to create a new user profile and retrieve their user ID. In a real application, this is where the authentication part of the application will take place. As previously mentioned, we will disregard all the password and authentication aspects of our application. Instead, we will treat the user ID as a secret number that the user will be issued. A new user is created by sending a request to the POST /users endpoint, passing the details

of the user profile as a JSON request body. The `UserService` service will then create the user and return it to the client.

2. **List the available books**: The user can then fetch a list of books. This corresponds to navigating the home page of the `BookSwap` application and viewing which books are available for swapping. The client issues a `GET /books` request. `BookService` will fetch a list of books, filtering them by available status.

3. **Swap a book**: Once the user has set up their profile and identified a book that they'd like to have, they can decide to swap a book. The client issues a `POST /books/{id}` request, passing the ID of the book that they want to book. Then, they pass their own user ID as a URL parameter, completing the URL of the request as `POST /books/id?user={userId}`. This could also have been implemented with a request body.

Looking at the main components in *Figure 6.2*, we notice that there is an extra component named `HTTP Server`. The implementation of `HTTP Server` in the BookSwap application consists of a few different parts:

- A handler custom type with **handler functions** that take in an HTTP request and response writer. We saw an example of a handler function with the `Index` example in *Chapter 5, Performing Integration Testing*. Typically, the handler custom type has access to all the dependencies it requires to fulfill its exposed operations.

- Each handler function is **configured to serve an endpoint** of the HTTP server. We use the `gorilla/mux` library to take care of the configuration and routing of requests to their respective handler functions. You can read more about the `gorilla/mux` library at `https://github.com/gorilla/mux`.

- Finally, once routes and handlers are set up, we start the server and configure it to listen on a given port. This is done using the `net/http` library in the Go standard library.

> **What is a mux?**
>
> **Mux** stands for **HTTP request multiplexer**. It provides functionality for receiving a variety of requests and routing them to the correct handler function based on the HTTP method, path, and query values. While there are other options, `gorilla/mux` is a popular solution among Go developers.

Figure 6.2 covers a successful journey for a new user. E2E tests should cover a variety of scenarios, so we map out multiple request flows. However, due to their higher cost and running time, they typically only cover **base cases** or **happy paths**.

Using Docker

Up until now, we have run applications using the `go run` command and tested them using the `go test` command. The downside of this approach is that we must set up the Go environment and any dependencies locally before we can build and run the code locally. In *Chapter 5, Performing Integration Testing*, we introduced Docker as a solution that addresses these issues.

> **What is a Dockerfile?**
>
> A Dockerfile is a file that contains all the commands required for assembling an image. It is then used by Docker Engine to automatically create and start a container. We can view it as the specification of the setup for a Docker container.

We need to create a custom Dockerfile for our `BookSwap` application, as it does not have a pre-defined image in the Docker Hub image library. The `Dockerfile.book-swap.chapter06` file contains the `BookSwap` specification:

```
FROM golang:1.19-alpine
WORKDIR /app
COPY go.mod ./
COPY go.sum ./
COPY . .
RUN go mod download
RUN go build ./chapter06/cmd
EXPOSE ${BOOKSWAP_PORT}
CMD [ "./cmd" ]
```

This relatively simple file demonstrates all the essential knowledge that we need to effectively use Docker with our Go applications:

1. The `FROM` statement indicates the base image of this build stage. We choose an image from Docker Hub that has the **1.19 Go version** that we require to run our application. In Docker terminology, `alpine` images are lightweight and run on the **Linux BusyBox distribution**.

2. The `WORKDIR` statement creates and sets the working directory of the Docker container. All further commands in our file execute in this directory.

3. Next, the `COPY` statement copies all the source files from our local directory to the container's working directory. Remember that containers are isolated from the underlying local directories, so these files must be copied to the container.

4. The `RUN` statement executes the commands needed to build our Go executable by first downloading its dependencies and then specifying the directory that contains our application

entry point. The Dockerfile is placed alongside the `go.mod` file in the root directory, so we need to explicitly state which of our chapter entry points to build from.

5. The `EXPOSE` statement instructs the Docker container to listen to network requests on the given port, as indicated by the `BOOKSWAP_PORT` environment variable. This variable is needed for the `BookSwap` application, so ensure that it is set in your terminal session before you run the application. The instructions of how to set environment variables will be different according to your operating system. If you want to run with the default setup, set the `BOOKSWAP_PORT` environment variable to `3000`.

6. Finally, the `CMD` statement specifies the command that should be run once the container starts. We run the executable from the `go build` step.

That's all we need to specify to run our application on any environment that is running Docker and has an internet connection! Most Docker specifications will be using this simple recipe for writing and running their custom images. We will see this specification in action later in this chapter.

Persistent storage

The next change we will make is to add **persistent storage** to the `BookSwap` application, allowing us to save the state once the application shuts down. As SQL databases are still the most popular persistent storage solutions, we will use a SQL database in the demo application.

> **SQL database management solutions**
>
> There are a few popular SQL database solutions that you may already be familiar with. Some of these are **Oracle MySQL**, **Microsoft SQL Server**, and **PostgreSQL**. They all allow us to manage and interact with underlying SQL databases but may have differences in terms of the way we interact with them. Therefore, it's important to use the same SQL solution in test environments as we have in production.

We will use a **PostgreSQL database**, an open source relational database widely used in production. It also has excellent cloud support across the public cloud providers, so it is a natural choice for our technology stack.

You can easily download and install it locally using the steps in the official documentation, available at `https://www.postgresql.org/download/`, following the instructions for your operating system. Once you install it locally, make a note of your host, port, username, and password. You will need these to connect to your database and run instructions on it later.

Figure 6.3 depicts the two main tables required for our application:

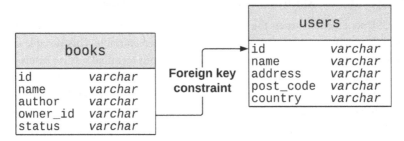

Figure 6.3 – The SQL tables of the BookSwap application

There are two tables: `books` and `users`. Each table has a column for the data fields of each type. As the `owner_id` value of `books` should belong to an existing user, there is a **foreign key constraint** between the two tables on the user `id` field. All the columns are of type `varchar`, which corresponds to a `string` type.

Before we start up the application, we need to create the tables required for the application to function. However, we don't want to execute SQL instructions outside of our source code, instead allowing us to version and review it.

The `golang-migrate` project (`https://github.com/golang-migrate/migrate`) allows us to write migrations and run them as part of our application startup. The `migrate` CLI tool can easily be installed on your environment by following the installation steps in the *Getting started* tutorials available in the project documentation.

Once installed, we can generate migrations for our two required tables:

```
$ migrate create -ext sql -dir chapter06/db/migrations -seq
create_users_table
$ migrate create -ext sql -dir chapter06/db/migrations -seq
create_books_table
```

The `migrate` CLI tool creates two files for each table, named `*.up.sql` and `*.down.sql`. The intended usage of these files is that the up migration creates the table at startup, and the down migration deletes the table once the application shuts down. This ensures that tables are removed after each run of the application and that they are always created at startup. Then, we add the columns required for each table. The configuration of the `users` table is shown here:

```
BEGIN;

CREATE TABLE IF NOT EXISTS users
(
    id VARCHAR (50) PRIMARY KEY,
```

```
   name VARCHAR (50) NOT NULL,
    // other column definitions
);
COMMIT;
```

The migration creates a table if one does not exist and marks the `id` column as the primary key. The specification matches the columns defined in *Figure 6.3*. The down migration is one single line, which drops the table:

```
DROP TABLE IF EXISTS users;
```

The configuration of the `books` table is done in the same way. Finally, we add the migrations to the entry point of the BookSwap application:

```
func main() {
// other initialisation code
   postgresURL, ok := os.LookupEnv("BOOKSWAP_DB_URL")
   if !ok {
      log.Fatal("env variable BOOKSWAP_DB_URL not found")
   }
   m, err := migrate.New("file://chapter06/db/migrations",
postgresURL)
   if err != nil {
      log.Fatal(err)
   }
   if err := m.Up(); err != nil {
      log.Fatal(err)
   }
   defer func() {
      m.Down()
   }()
   // other initialisation code
}
```

Interacting with the `migrate` library requires three extra steps to be added to the startup of the application:

1. Creating a **new migrate instance** with a source URL and database URL. As previously mentioned, the connection string for your database instance will depend on your SQL configuration. The format for your database will have a format similar to `postgres://user:password@`

host:port/database. The BookSwap application requires an environment variable named BOOKSWAP_DB_URL, which contains this value. Make sure that this environment variable is set before starting the application.

2.   Once the migrate instance is created, we invoke the Up() method. This method looks at the current migration version and applies all the migrations defined in our *.up.sql files.

3.   If we want to clean up after the application shuts down or in the case of an error, we defer a call to the migration Down() method. This method uses the same version management and runs the contents of the *.down.sql files.

The golang-migrate library and CLI have made it easy for us to save database configuration alongside source code, allowing for easy versioning as well as version management.

Once our database and tables are created, we need to refactor our UserService and BookService implementations to make use of SQL tables and not the built-in maps we have been using so far. Typically, engineers make use of an **object-relational mapping** (**ORM**) library, which allows us to create a bridge between our Go custom types and our PostgreSQL database.

There are a few ORM solutions in the Go ecosystem. One of the most popular ones is **GORM** (https://github.com/go-gorm/gorm), which is an open source easy-to-use Go library. This library will make it easy for us to interact with our database solution, removing the need to manage SQL as raw strings in our source code.

The setup for using GORM is very similar to what we have done using golang-migrate:

```go
func main() {
  // other initialisation code
  postgresURL, ok := os.LookupEnv("BOOKSWAP_DB_URL")
  if !ok {
    log.Fatal("$BOOKSWAP_DB_URL not found")
  }
  dbConn, err := gorm.Open(postgres.Open(postgresURL), &gorm.Config{})
  if err != nil {
    log.Fatal(err)
  }
  ps := db.NewPostingService()
  b := db.NewBookService(dbConn, ps)
  u := db.NewUserService(dbConn, b)
  // initialisation code continues
}
```

First, we connect to the database using the database URL that we used previously to connect with `golang-migrate`. Once successfully connected, it returns the GORM database wrapper of type `*gorm.DB`.

If we cannot connect to the database, then we will kill the application. We have also changed the signatures of the `NewBookService` and `NewUserService` initialization functions to take in the initialized database session.

All of the operations of `BookService` and `UserService` that previously saved models to the built-in map type will now have to leverage the operations of the GORM database wrapper. One example is the `BookService` `ListByUser` method:

```
// ListByUser returns the list of books for a given user.
func (bs *BookService) ListByUser(userID string) ([]Book,
error) {
   var items []Book
   if result := bs.DB.Where("owner_id = ?", userID).
Find(&items);
      result.Error != nil {
         return nil, result.Error
   }
return items, nil
}
```

This method lists all books whose owner ID matches the given user ID. The usage of an ORM solution allows us to invoke operations on the database using a service that is able to use the invocations of methods to construct the correct SQL query. This allows us to reduce errors and remove the need to manage and interact with raw SQL strings.

Running the BookSwap application

The final piece of the puzzle for extensions we need to make to the `BookSwap` application is how to run it together with its database. As mentioned in *Chapter 5*, *Performing Integration Testing*, the `docker-compose` tool is what we use to manage multiple Docker containers together. We now have two services, or parts, to the `BookSwap` application—the server-side application and its database. The database setup needs to be run in a Docker container as well, to avoid having to set up databases in every single environment.

The docker-compose command takes its input from a .yml file, which makes it easy to specify the different services and their requirements. This simple specification in our docker-compose. book-swap.chapter06.yml file defines our BookSwap application and a PostgreSQL database that it can use:

```
version: '3'
services:
 books:
  build:
   context: .
   dockerfile: Dockerfile.book-swap.chapter06
  ports:
   - "${BOOKSWAP_PORT}:${BOOKSWAP_PORT}"
  depends_on:
   db:
     condition: service_healthy
  restart: on-failure
  env_file:
   - docker.env db:
  image: postgres:15.0-alpine
  ports:
   - "5432:5432"
  expose:
   - "5432"
  env_file:
   - docker.env
  restart: on-failure
```

The configuration in this relatively simple file specifies everything we need:

1. We define a services block for all the services that we start up. In our case, we will define the books service and the db service, each configured in its own sub-block.

2. The configuration of the books service specifies the following:

    I.   The service is built from its own Dockerfile located in the current directory. This is the Dockerfile that we previously discussed in the *Using Docker* section of this chapter.

    II. The service exposes the port specified by the BOOKSWAP_PORT environment variable on its network. This will allow us to run tests that need access to the port locally.

III.   The service depends on the db service being started successfully. Docker Engine will take this into account when starting our service and instruct that the db service be started first.

IV.   The service uses the `docker.env` file for its environment variable configuration. This will specify other environment variables that we require, like `BOOKSWAP_DB_URL` that we have previously seen.

3.   The configuration of the db service specifies the following:

I.   The container should use an existing image from the Docker repository. At the time of writing, this was the latest image of the **PostgreSQL** image, available at `https://hub.docker.com/_/postgres`.

II.   The service exposes port `5432`, as is customary for PostgreSQL.

III.   As can be seen from the documentation for this image, it requires the specification of a number of environment variables for the database username, password, and name. All of these variables will be defined in the `docker.env` file that we supply to this service.

4.   Both services have a restart policy defined as well. This means that Docker will automatically restart containers if they fail.

We can provide the following sample configuration for `docker.env`, but you can easily change it according to your own preferences by editing the file:

```
POSTGRES_USER=root
POSTGRES_PASSWORD=root
POSTGRES_DB=books
BOOKSWAP_DB_URL=postgres://root:root@db:5432/
books?sslmode=disable
BOOKSWAP_PORT=3000
```

That's all the configuration required for running two services together with the `docker compose -f docker-compose.book-swap.chapter06.yml up --build` command from the project root directory. This file contains a typical configuration that you will be able to reuse in your own projects. Furthermore, it allows us to identically start and run the BookSwap application across different environments. This provides us with the key advantage of being able to easily spin up test environments for the application as a whole. With these building blocks in place, let us have a look at how we can make use of this key advantage to increase the test coverage of our application.

Exploring Godog

In this chapter, we have made quite a few changes that have extended the scope and complexity of the `BookSwap` application. Now that we can easily start and tear down the application using Docker containers, it is time to turn our attention to writing E2E tests for our application.

In *Chapter 5, Performing Integration Testing*, we looked at how to write BDD-style tests. This style of testing allows us to write human-readable test scenarios and use a **Given-When-Then** structure. These readable tests can serve as documentation for our projects, allowing us to involve multiple stakeholders and write tests that truly cover the functionality of our applications.

We also explored the `ginkgo` testing library, which allowed us to write tests using this style. Godog (`https://github.com/cucumber/godog`) is another testing library that we will be exploring to write BDD-style tests. `ginkgo` allows us to add BDD-style assertions to our unit tests, but Godog provides extra code generation capabilities that make it a great fit for writing functional tests. We will learn how to use this great library for integration and E2E testing.

Here are some of the highlights of Godog:

- Unlike the libraries we have used so far, Godog does not run its tests using the `go test` command, but with the `godog run` command. This command serves the dual purpose of generating test files as well as running the test that has been implemented.

- Tests are organized in **feature files**, which describe the expected behavior of a particular piece of functionality in a particular scenario. Godog uses a domain-specific language called **Gherkin** (`https://cucumber.io/docs/gherkin/reference/`). We will explore how to write tests in this format for the remainder of this chapter.

- Godog is an open source library, maintained by the community and the Cucumber organization. You can freely explore the source code and even contribute.

Just as with the other dependencies we have used so far, Godog is installed by running the `go install` command in the terminal:

```
$ go install github.com/cucumber/godog/cmd/godog@latest
```

Now that we understand the basic usage of Godog and have installed it successfully, we start by writing our first **feature file**. We start out with a simple feature file for the `BookSwap` application:

```
Feature: New user signs up
  In order to use the BookSwap application
  As a new user
  I need to be able to sign up.
Background: Verify configuration
  Given the BookSwap app is up
```

```
Scenario: Sign up
  Given user details
  When sent to the users endpoint
  Then a new user profile is created
```

The feature file describes a part of the functionality required for new users of the BookSwap application:

- The feature describes the scenario of signing up as a new user of an application.

- As a background step, the BookSwap application should be up and running. This allows us to write an E2E test as we run the entire application and run the test on the side.

- When the feature is completed, the following functionality will be available:

  - New customers will be able to create user profiles.

  - When their profile is created, the user will see their user summary and receive their user ID, which will allow them to further interact with the application.

  - Once signed up, customers will be able to view their profile by using their user ID.

  - Any further interactions with the application are outside the scope of this feature.

As discussed, the feature file is based on the expected user journeys and request flows of the application. Feature files should be easy to read and understand, so we should create separate files for covering other features and scenarios and use non-technical language.

In the next section, we will learn how to implement and run this feature file with Godog.

Implementing tests with Godog

With Godog installed and our first feature outlined, let us turn our attention to the implementation of this test.

The main steps we will be taking for implementing our outlined feature are as follows:

1. Creating feature and test files.

2. Implementing the test steps for the functionality of our BookSwap application.

3. Running the application as well as the test.

As we previously mentioned, we will use Godog to implement BDD-style E2E tests, so we require the application to be up and running before we run our tests. However, this is not a Godog requirement, so we can write tests at any level with this easy-to-use library.

Creating test files

As previously mentioned, Godog relies on code generation to make developers' lives easier. The process consists of copying code from the terminal and creating files ourselves. Let us look at the steps involved.

Step 1 – creating a feature file

Feature files are stored in the /features directory located at the root of the Go project. As we use project folders in our repository, we need to create a file under /chapter06/features. We will create a file in this directory and add the feature text inside it:

```
$ mkdir chapter06/features
$ vim chapter06/features/newUserSignsUp.feature
```

Note that the file is named according to the feature name, making it easy to understand which functionality the file relates to.

Step 2 – generating step definitions

Once the feature file contains our text, we can use Godog to generate the steps required for our feature. The godog run command prints the following generated code to the terminal:

```
func aNewUserProfileIsCreated() error {
    return godog.ErrPending
}
func sentToTheUsersEndpoint() error {
    return godog.ErrPending
}
func theBookSwapAppIsUp() error {
    return godog.ErrPending
}
func userDetails() error {
    return godog.ErrPending
}

func InitializeScenario(ctx *godog.ScenarioContext) {
    ctx.Step(`^a new user profile is create,
        aNewUserProfileIsCreated)
    ctx.Step(`^sent to the users endpoint$`,
        sentToTheUsersEndpoint)
    ctx.Step(`^the BookSwap app is up$`, theBookSwapAppIsUp)
```

```
    ctx.Step(`^user details$`, userDetails)
}
```

Figure 6.4 presents the sequence of steps in our scenario, along with any HTTP requests that they make:

	Name	Request
1	aNewUserProfileIsCreated	GET /users/{id}
2	sentToTheUsersEndpoint	POST /users
3	theBookSwapAppIsUp	GET /
4	userDetails	

Figure 6.4 – Steps and HTTP requests made in our scenario

The generated code contains a function for each step of our scenario:

1. The aNewUserProfileIsCreated function sends a request to the GET /users/{id} endpoint and verifies that the user profile is successfully created. It will also verify that the user profile can be successfully retrieved by using the assigned user ID.

2. The sentToTheUsersEndpoint function sends a JSON payload to the POST /users endpoint and verifies that the endpoint responds with the correct user details. It will also get access to the user ID that the application generates for the new user profile.

3. The theBookSwapAppIsUp function sends a request to the GET / endpoint and verifies that the application responds with a 200 OK status code. In production, we often expose a separate /health endpoint, but we will make use of the root endpoint for the purposes of our BookSwap demo application.

4. The userDetails function will create a db.User instance that we will marshal to the JSON payload and send to the sentToTheUsersEndpoint step. It will also serve as the expected value, or want variable, in our test assertions.

We will need to implement these functions to invoke the functionality of our application.

Finally, the `InitializeScenario` function ties together all these functions into steps and orders them alphabetically. We will have to correctly order them according to our feature definition when we implement our test file.

While the generated code is simple, it provides a scaffold for our test code and takes care of the interaction required with the Godog test runner.

Step 3 – creating a test file

Just as with regular unit tests, Godog tests also live in the `*_test.go` files and live alongside the packages that they test. As we will test the entire application, we create a test file in the `root` directory, at the same level as the `/features` directory. We create a test file matching the name of the feature and paste the generated file inside it:

```
$ vi /cmd/newUserAddsBook_test.go
```

While the name of the test does not need to match, using a matching test will allow Godog to match the test with the feature.

With the test file and code created, we execute `godog run` again. The test runner will mark the scenario as `pending`:

```
Background: Verify configuration
  Given the BookSwap app is up # newUserSignsUp_test.go:148 ->
theBookSwapAppIsUp
      TODO: write pending definition


Scenario: Sign up        # features/newUserSignsUp.feature:9
  Given user details      # newUserSignsUp_test.go:152 ->
userDetails
  When sent to the users endpoint # newUserSignsUp_test.go:144
-> sentToTheUsersEndpoint
  Then a new user profile is created # newUserSignsUp_test.
go:140 -> aNewUserProfileIsCreate
```

Conveniently, the output prints out the line numbers for each step as well, showing us where we are missing the implementation required for our tests.

Implementing test steps

Now that Godog has conveniently generated a scaffold for our test steps, we begin writing test code according to the functionality of the `BookSwap` application. However, as described in the previous section, we will need to pass information between test steps.

The way to do this in Godog is by passing information through chained contexts. Godog will pass the contexts between the test steps, allowing us to pass information between steps in a safe way. In order to do this, we will need to change the signature of the test steps to take in a context and return a context and an error:

```
func theBookSwapAppIsUp(ctx context.Context) (context.Context,
error) {
    // test step implementation
}
```

The test steps take in a context and return a context and error. Under the hood, Godog will handle each of these return values correctly: chaining the returned context to subsequent test steps and failing the test in the case of a non-nil error.

> **Context refresher**
>
> The context type is part of Go's standard library and its purpose is to carry deadlines, cancellations, and request-scoped variables. Contexts should be propagated across functions, allowing us to tie together function calls to requests across the layers of our application. Creating a new context requires a parent context. Cancellations then propagate across the chain of children contexts.

For our purposes, we will use contexts to carry request-scoped variables. We will create a new contextKey custom type that will carry all the variables that we need to pass between test steps:

```
// contextKey is used to pass information between test
// steps.
type contextKey struct {
    UsersURL string
    User     db.User
}
```

In our case, we will propagate the BookSwap application's UsersURL and the wanted value of the created user. In our background step, theBookSwapAppIsUp, we see a demonstration of the usage of the context to pass information to subsequent steps:

```
func theBookSwapAppIsUp(ctx context.Context) (context.Context,
error) {
    url, err := getTestURL()
    if err != nil {
        return ctx, fmt.Errorf("incorrect config:%v", err)
    }
```

```
  resp, err := http.Get(url)
  if err != nil || resp.StatusCode != http.StatusOK {
    return ctx, fmt.Errorf("bookswap not up:%v", err)
  }
  return context.WithValue(ctx, contextKey{}, contextKey{
    UsersURL: url + "/users",
  }), nil
}
```

This code snippet demonstrates the implementation of a step that interacts with a REST endpoint:

1. We set the URL value for the environment that we will be testing by calling the `getTestURL` helper function. This function constructs the URL based on the environment variables specified for the application. This makes it easy for us to configure our test to run in different test environments, local or remote. If you want to run with the default values, set the `BOOKSWAP_BASE_URL` environment variable to `http://localhost` and the `BOOKSWAP_PORT` environment variable to `3000` to your terminal session.

2. We use the `http.Get` method to interact with the defined URL, saving the error and the response. We are familiar with the `net/http` library from previous chapters. Its usage is no different in this test.

3. In the case of an error or a status code other than `200 OK`, we return an error. This will fail this step and end the test.

4. Finally, in the case of success, we use the `context.WithValue` function to create a child context from the `ctx` parameter value, passing a `contextKey` value with the populated `UsersURL`. In later steps, we will be able to use this URL for our requests.

One other change we need to make to the generated test steps is to reorder them according to the order in which they should run. This step is not intuitive if you have never used Godog, but will be easy to track down if forgotten as your test will fail.

Running the test suite

With everything implemented, it's time to take our test out for a spin. First, we remember to run the BookSwap application, either using `docker compose -f docker-compose.book-swap.chapter06.yml up --build` command. Unless you have changed your configuration, this will expose the application at the `http://localhost:3000` URL. You can easily verify that the application is running by performing a `curl` command against this endpoint:

```
$ curl --location --request GET 'http://localhost:3000'
{"message":"Welcome to the BookSwap service!"}
```

If you see a welcome response, then the application is up and correctly connected to its database.

With the application running, we execute our test using the `godog run` command:

```
$ cd chapter06 && godog run
Feature: New user signs up
  In order to use the BookSwap application
  As a new user
  I need to be able to sign up.
Background: Verify configuration
  Given the BookSwap app is up  # newUserSignsUp_test.go:23 ->
theBookSwapAppIsUp
Scenario: Sign up                # features/newUserSignsUp.
feature:9
  Given user details      # newUserSignsUp_test.go:35 ->
userDetails
  When sent to the users endpoint # newUserSignsUp_test.go:50
-> sentToTheUsersEndpoint
  Then a new user profile is created # newUserSignsUp_test.
go:84 -> aNewUserProfileIsCreated
1 scenarios (1 passed)
4 steps (4 passed)
11.876996ms
```

As we can see from the terminal output, Godog runs one scenario and all four steps passed. Alternatively, you can run the tests using the `go test` command if you don't want to install the Godog CLI, but that will not format the test results, as you see in the preceding output.

We have successfully written and run our first E2E test for the `BookSwap` application, which has been extended with persistent storage. The test was written using the Godog open source testing library, which allowed us to write easy-to-read BDD-style tests. We are well on our way to becoming Go testing experts.

Using database assertions

We have learned how to start up our application in a test environment, as well as how to write and run E2E tests for our application. This has taken us far toward verifying the behavior of our application, but how can we be sure that the stored data and database components are correct? The final aspect of E2E testing that will help us answer this question is database testing. Looking at the tests we have written so far, we notice two things:

- The database is typically initialized as empty, then the tables are torn down once the application shuts down. This has the advantage that we know there will be no persistent data that interferes with our tests, but it has the disadvantage of having to set up any required data as part of the test. For example, in our case, registering an available book requires a user ID, so we will have to create a user first before we do any book-related tasks. This can make our test suite running time longer.

- The items from the database are asserted through the `BookSwap` endpoints. For example, we check that a user was correctly stored in the database by making a request to GET `/users/{id}` with its corresponding user ID. However, as requests travel down the entire application stack, it can become difficult to pinpoint the source of the error.

Let us explore these two pain points further to get a better understanding of how to address them.

Seed data

As discussed in the previous *Persistent storage* section, we use `golang-migrate` to write database migrations, which allows us to create and set up our database for usage before the application starts up. Then, we use the **GORM** library as our ORM, which allows us to easily interact with our database using custom types.

The next step would be to insert some data into our newly created tables. This type of initial data is known as **seed data**. We can add seed data to our databases by simply adding the corresponding SQL statements to the migration file. For example, we can create an initial user after the table is created in the up migration by using the `INSERT` command:

```
INSERT INTO users VALUES ('ABC-123','Initial user', '1 London
Road', 'N1', 'UK');
```

However, none of the tools we have explored have the capability of creating and adding random seeds for our application. We could add another library dependency for generating random data, but instead, we can make use of GORM's DB type to insert random data into our database before our tests run:

```
func addUser() error {
  dbConn, err := gorm.Open(postgres.Open(postgresURL), &gorm.
Config{})
  if err != nil {
    return err
  }
  dbConn.Save(&db.User{
    ID:        uuid.New().String(),
    Name:      "Generated User",
  })
```

```
   return nil
}
```

The previous code snippet demonstrates how to insert data into the database alongside running our Godog test steps, as detailed here:

1. Just as at the application start, we open a new connection to our PostgreSQL database. Database connections should be shared between tests as much as possible, and you should not open them for many test iterations. However, as database start positions are usually required for E2E tests, it's feasible to set up our database in this way.

2. Once the database connection is opened successfully, we save a generated user using the `Save` method of the GORM database. After this, the database will contain the generated data and can be used across our tests.

Test cases and assertions

When it comes to assertions on the contents of our database, we can take the same approach as we've seen with adding generated data:

```
func verifyUser(want db.User) error {
   dbConn, err := gorm.Open(postgres.Open(postgresURL), &gorm.
Config{})
   if err != nil {
     log.Fatal(err)
   }
   var got db.User
   if err := dbConn.Where("id = ?", want.ID).First(&got);
   err != nil {
     return err.Error
   }
   if want != got {
     return fmt.Errorf("user does not match:got %v, want
%v",got, want)
   }
   return nil
}
```

Looking at the verification code, we see the same approach:

1.   We connect to the database using the connection string. GORM will optimize the use of your database connection when being used by multiple goroutines.

2.   Then, we use the database methods to query the database for the user ID that is supplied to the method. Note that we rely on the GORM database directly, and not our own methods of `UserService`, removing any possible bugs that we may have introduced.

While GORM is easy to use, it might feel quite verbose to write full database queries to assert on returned values. The `dbassert` open source library (`https://github.com/hashicorp/dbassert`) provides wrappers and helper functions that can make this easier for you. You can explore this library by yourself and see how it can help streamline your test code.

Summary

In this chapter, we spent our efforts extending the `BookSwap` application. We began by discussing what a typical user journey for a user will be, added a PostgreSQL database to it, and configured it to run with Docker. Then, we explored the Godog testing library, which makes it easy to write BDD-style tests, as well as E2E tests. We made use of Godog to verify that users are able to sign up on the `BookSwap` application, making use of the code generation abilities of Godog. Finally, we briefly discussed the challenges of creating database start positions and assertions directly at the database level.

In *Chapter 7*, *Refactoring in Go*, we will discuss tools and techniques for code refactoring and how to break up monoliths into multiple services. This will give us a realistic understanding of how to rely on our tests to verify that refactoring does not cause errors or break existing functionality.

Questions

1.   What is the purpose of a user journey?

2.   What is ORM?

3.   What is the advantage of using Docker Compose?

4.   What is database seeding?

Further reading

- *User Story Mapping: Discover the Whole Story, Build the Right Product, Jeff Patton, Peter Economy.* Published by *O'Reilly*.

- *The Book Of Kubernetes: A Hands-on Deep Dive into Container Technology, Alan Hohn.* Published by *No Starch Press*.

- *SQL & NoSQL Databases: Models, Languages, Consistency Options and Architectures for Big Data Management, Andreas Meier, Michael Kaufmann.* Published by *Springer Vieweg*.

7
Refactoring in Go

We have explored concepts and techniques for tests across the entire testing pyramid. We have applied these concepts while building our main project of study, the BookSwap application. This web application is currently verified by the following:

- Unit tests implemented using the Go standard library
- Integration tests implemented using httptest
- End-to-end tests implemented using godog

To demonstrate these techniques in a realistic example, we have extended the functionality of the BookSwap application with a variety of components. In *Chapter 6, End-To-End Testing the BookSwap Web Application*, we extended the project by adding the ability to run it in Docker and use a PostgreSQL database to save its data.

All of these changes have added complexity to our BookSwap application, which now relies on the following:

- Two libraries for database migration and operations – golang-migrate and gorm
- Three different types of files – source files, implementation files, and Docker files
- A complex code structure with multiple layers – db, handlers, and cmd

The BookSwap application started as a simple REST API with a small scope of functionality. However, as we kept refining it and adding more code, it became more difficult to install and start up. This is part of the natural life cycle of software projects. As an engineer, you will more often have to modify and extend existing code, performing **brownfield development**, than start and implement brand-new projects with no existing dependencies, also known as **greenfield development**.

This chapter is dedicated to **code refactoring**, which is modifying and restructuring existing code. Based on our experiences of implementing and testing the `BookSwap` application, we will learn about good practices for code refactoring. Then, we will learn how to validate the behavior of the restructured code, which should perform and behave identically to its legacy equivalent. Finally, we will discuss best practices for splitting up monolithic applications into microservices.

In this chapter, we will cover the following topics:

- What code refactoring is and why it is an essential part of the development process
- How to effectively change implementation and test code
- Error verification in Go
- How to validate refactoring success criteria
- Good practices for splitting up monolithic applications

Technical requirements

You will need to have **Go version 1.19** or later installed to run the code samples in this chapter. The installation process is described on the official Go documentation at `https://go.dev/doc/install`.

The code examples included in this book are publicly available at `https://github.com/PacktPublishing/Test-Driven-Development-in-Go/chapter07`.

Understanding changing dependencies

In *Chapter 1*, *Getting to Grips with Test-Driven Development*, we discussed refactoring the code we are writing as part of the **Red-Green-Refactor** TDD technique. This involved limiting the cleaning up of code as we write it. However, as we continue our journey with TDD, it is essential that we consider how our code will evolve through time and consider larger-scale code refactoring or rewrites.

Code refactoring is often used interchangeably with **code redesigning**, but they represent different levels of code modification. Code redesigning involves changing the functionality of a code base/service, while code refactoring involves changing the way the service delivers its existing functionality. In fact, if done correctly, code refactoring will be invisible to any internal and external users of the service functionality.

> **The purpose of code refactoring**
>
> Developers refactor their code to make it more efficient, maintainable, and extendable. There are many benefits to code refactoring: better readability, improved performance, and enabling developers to change the code more efficiently. Together, these are known as **non-functional requirements**.

A project's testing strategy is an essential aid for verifying and supporting efficient code changes and will help developers avoid the following:

- **Functionality regressions**: The refactored code should not break any existing functionality, causing a regression. Integration tests will identify components that might no longer work together correctly, while end-to-end tests will pinpoint which breakages affect user-facing functionality.

- **Performance degradation**: The refactored code should not perform any slower than the existing functionality. Integration tests will identify which components have slowed down for a particular scenario or operation, signaling to developers which components should be investigated further. End-to-end tests will identify which performance issues are affecting users but might make it more difficult to isolate the problem, as they don't provide the granularity of system components. However, they will give an important indication of the severity of a particular performance issue, allowing developers to correctly prioritize issues. We will cover performance testing in more detail in *Chapter 8, Testing Microservice Architectures*.

- **Changes outside the intended scope**: The refactored code should not affect components outside the intended scope of the changes. This indication is particularly important for legacy code bases, where developers might not have a clear picture of the dependency graph of the different components. Unit tests will pinpoint which packages within the current codebase/ service might be affected by the refactor, while integration tests will highlight whether the APIs between different services might be broken.

The potential costs of these issues come in multiple forms:

- Losing business/transaction volume during a potential outage

- Increased infrastructure/cloud costs in the case of slower performance

- Increased development costs if developers take a longer time to deliver a code change

Therefore, it is essential that code refactoring is easy to take on and verify.

Code refactoring steps and techniques

Now that we understand the fundamental need for code refactoring, let us explore some code refactoring techniques. These are not limited to Go development itself, but it is important to understand the process by which we change the code so that we can then understand how to effectively validate its output. *Figure 7.1* depicts the basic working process of code refactoring:

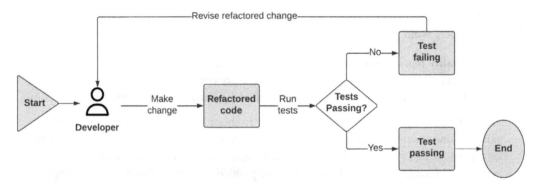

Figure 7.1 – The working process of code refactoring

The code refactoring steps rely on tests for verification:

1. The developer identifies the change that they want to make to the existing implementation.

2. Then, they make the required change, ensuring that the code continues to compile. This might require making changes to both the implementation and testing code.

3. Once their first change is done, the developer runs the test suite to verify their implementation changes.

4. If the tests are passing, then this refactor is successful and the developer has successfully implemented this change. They can proceed to commit it and release it.

5. If the tests are failing, then this refactor is not successful and the developer must revise their refactored change. This might mean making further changes to either the implementation or test code or simply adjusting their new code changes.

The working process is closely related to the red-green-refactor process that we have seen in earlier chapters. The developer should not share any changes that they make without the test suite successfully passing. This is most often enforced by commit checks and test run verifications as part of the build/release pipeline.

As depicted in *Figure 7.2*, code refactoring should consist of a series of minor code changes or modifications, ensuring that the refactored code retains the same major functionality:

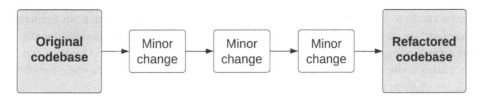

Figure 7.2 – Refactoring as a series of minor changes

As with many aspects of the code development process, releasing small, isolated changes is better than large code releases. This allows developers to verify each code change and release it in turn. Furthermore, in the case of issues, reverting a small code change will be easier than reverting larger code changes that have been committed over multiple days.

Figure 7.3 depicts an overview of five popular code refactoring techniques:

Technique Name	Main approach	Test changes?
Red-green-refactor	Three systematic steps that involves writing a **Red** test, writing the functionality required to make the test **Green** and then **Refactoring** the initial implementation.	Most likely
Extract	**Extract** an existing code fragment from its original function to a newly established function, which describes its purpose.	Not likely
Simplify	**Simplify** the logic of an existing function by consolidating its conditional expressions and adjusting the function parameters	Most likely
Inline	Remove redundant functions and their dependencies by removing their code and **inlining** them to its callers.	Not likely
Abstraction	Create new levels of **abstraction** such as interfaces to remove repetition and complex method signatures.	Most likely

Figure 7.3 – Five popular code refactoring techniques

The five techniques can be used together to improve code complexity and duplication:

- **Red-green-refactor** is the technique we are already familiar with. The implementation is written alongside its corresponding tests, starting with a failing test, making it pass, and then refactoring the written code as required. This approach ensures that all functionality is covered by tests and that the refactoring is undertaken as part of the initial implementation. As tests are written alongside the code, this technique will most likely require test changes as part of the code refactoring process.

- **Extract** is the technique that involves extracting an existing code fragment from a potentially large function into its own function. This function name should describe the functionality that the extracted fragment implements, improving the readability of the previous large function containing multiple pieces of functionality. As code is only extracted, not rewritten, test changes will not likely be required.

- **Simplify** is the technique that improves the complexity of large functions. This can be done by refactoring conditional expressions or adjusting method calls by refactoring function parameters or adjusting interface signatures. As this technique involves changing function signatures, test changes will most likely be required.

- **Inline** is the opposite technique to **Extract**. It involves removing redundant functions by taking their contents and putting them in place of the existing function call. This reduces the indirection of the code, reducing the cognitive burden of the developer reading the code. Unless the method being tested is removed, this technique will not likely require test changes.

- **Abstraction** is the technique most likely suited for larger-scope code refactoring. This technique involves introducing new levels of abstraction, such as interfaces, to remove repetition and allow the reuse of behaviors across multiple packages. Since new interfaces will require the use of mocks and larger scope refactoring, this technique will most likely require test changes.

These popular techniques will help you refactor your code and ensure that it continues to adhere to the SOLID principles we have previously discussed in *Chapter 3, Mocking and Assertion Frameworks*.

Technical debt

Code refactoring is an extremely important and unavoidable part of the development life cycle. When code is not routinely refactored and maintained, it begins to accumulate **technical debt**. The subject of how to effectively manage technical debt has been discussed often in the engineering community, as it is easy for engineering managers to prioritize delivering new features, which have a tangible monetary value, as opposed to addressing technical debt, which does not have immediate consequences or cost associated with it.

> **Technical debt in Agile**
>
> In Agile, technical debt is the term used to refer to the consequences of prioritizing speed over quality. While the code is tested for correct functionality, its internal structure might be the result of poor architectural choices that have been made for speed.

The consequences of technical debt can affect your project in a variety of ways:

- **Bugs**: As code accumulates technical debt, duplicated code and high cohesion can lead to bugs that are difficult to fix and detect. These can have financial consequences in the case that they cause outages or affect user operations.

- **Decreased productivity**: As technical debt does not follow SOLID principles and does not resemble the rest of the code base, it can be difficult for developers to change it with new features. Furthermore, technical debt is also typically not well documented, so it can be difficult to reason around expected behavior.

- **Limits new features**: As it accumulates, developers can spend their entire time fixing bugs and performance issues with technical debt, meaning that they do not have time to deliver new features. This "putting out fires" and chaotic way of working frustrates engineers and can even lead to higher staff attrition.

Technical debt is often compared to financial debt. If we don't take care of the issues in our code for a prolonged period of time and continue extending code that has been poorly designed, the debt becomes larger and more difficult to handle, similar to how financial debt accumulates interest. To avoid these consequences, technical debt should be handled alongside other work in the Agile way of working.

Figure 7.4 depicts how Agile teams typically structure their work to address technical debt with a prioritized work **backlog**:

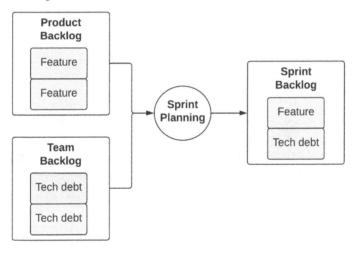

Figure 7.4 – Sprint planning with technical debt

The **sprint backlog** is a combination of feature work and technical debt:

- The development team and product team each maintain their own backlogs. Typically, these are represented by Jira tickets or GitHub issues with details of the work to be done. The technical debt work will typically involve refactoring existing code, while the feature work will consist of adding new functionality. The sprint backlog attempts to find a balance between these two types of work.

- During sprint planning, representative stakeholders prioritize the work. It is considered good practice to involve the development team during planning to ensure that the entire team has a good understanding of the goals of the upcoming sprint. The engineering team consists of experts who can scope what refactoring work should be taken on. They often have an understanding of which parts of the system require attention.

- The outcome of the sprint planning meeting is a prioritized list of work that makes up the sprint backlog. Based on their expertise, the technical team typically provides time estimates for the work to be done. These estimates are then used to determine what work can be undertaken to match the capacity of the team. Refactoring and feature work are treated as equal, with time being given to each piece of work based on provided estimates.

While technical debt seems like it does not have any immediate cost, it's important that teams are allowed to take time to refactor and maintain their code.

> **Planning for technical debt**
>
> A "little and often" approach works best, where technical debt is planned alongside feature work, as opposed to a "Big Bang" approach, where a whole sprint is dedicated to fixing issues and doing extensive code refactors.

Changing dependencies

Now that we have a good understanding of how to plan and undertake code refactoring, it is time to turn our attention to a special case of code refactoring, which is changing dependencies. As discussed in *Chapter 3, Mocking and Assertion Frameworks*, dependencies are typically wrapped in our own interfaces. Go's powerful interface also helps us refactor our code when dependencies change.

Interfaces make our code easier to refactor, as well as less coupled, by providing the following:

- **Clear expected behavior**: Interfaces are defined on the calling side, establishing what the expected behavior of the external dependency will be inside the package. Developers have a clear way to indicate what functionality and method signatures they expect the external dependency to provide.

- **Compiler-enforced method signatures**: Once the expected behavior is defined, the compiler will verify that any struct that is passed as the interface will satisfy these signatures. Therefore, the code can never get into a state where a certain method is not defined, and cause runtime errors once called.

- **Separation between packages**: As the interface lives inside the calling package, it provides a barrier between the package and the external dependency. The dependency can then be refactored or have new functionality implemented without the need to handle these changes in the calling packages.

- **Implementation opacity**: The calling package does not have any knowledge of the external dependency. This makes it easy for us to replace one concrete implementation with another, which makes changing dependencies during refactoring easy. We have already seen an example of this in *Chapter 3, Mocking and Assertion Frameworks*, where we saw how to provide mocks for dependencies of the UUT, allowing us to test its behavior in isolation from all its other concrete dependencies.

A good example to analyze this is `PostingService` of the `BookSwap` application. The purpose of the service is to take on the order and provide all the book shipping functionality:

```
// PostingService interface wraps around external posting
// functionality.
type PostingService interface {
  NewOrder(b Book) error
}
```

As expected, we have defined `PostingService` as an interface that we have defined inside the BookSwap application. This service has not been implemented, as we have considered the implementation of this service as fully external. Please note that this is a purely fictional service that we have used to demonstrate the process of providing and consuming external dependencies.

`BookService` takes `PostingService` using this defined interface as a dependency:

```
// NewBookService initialises a BookService given its
// dependencies.
func NewBookService(db *gorm.DB, ps PostingService)
*BookService {
  return &BookService{
    DB: db,
    ps: ps,
  }
}
```

We can provide any implementation of this function as long as it provides the method defined by the interface. This makes it easy for us to provide any implementation of this service to the method, without making any further code changes inside this package, which means we can keep the scope of the refactoring small to the implementing package.

> **Nil values as dependencies**
>
> The zero value of an interface is nil, so it will satisfy the method signature when passed in as well. While a nil value dependency can cause errors if a function is called on it, using nil for dependencies we are not interested in is particularly useful when writing tests.

While implementation is easy to swap when the signature of the dependency stays as expected, changing the interface method signature is not so easy and will require us to make changes to the calling packages, which have defined their own wrapping interfaces.

When changing the signature of an interface, you will typically need to make the following changes:

1. Make the changes to the implementation of the interface, if it is part of your project.

2. Update the test code to ensure that the refactored changes are working correctly. This will ensure that you have not introduced any bugs or caused regressions.

3. Based on the compiler errors, you can easily identify which packages use the implementation type as a dependency, as they will no longer satisfy these old method signatures. Then, you can make the corresponding changes to any interfaces that wrap around the implementation.

4. If you are using a mock generation tool, you can regenerate your mocks according to the newly updated interface definitions.

5. As identified by compilation errors, you can make any changes to the test code. These changes might be required after regenerating the mocks or to test the new behavior of the refactoring.

> **The compiler is your guide**
>
> The enforcement of interface signatures will help you identify which packages must be modified and ensure that the code does not end up in an unstable state. The compiler will highlight any code that needs to be modified and guide you in your refactoring efforts.

As changing interfaces requires a lot of rework, developers will try to avoid making these changes. However, taking the time to design your code according to good architectural principles should help you avoid needing to make such sweeping code changes often.

Relying on your tests

We now have a good understanding of how to refactor our code and have learned how to take advantage of some of Go's best features: the compiler and interfaces. This should make your refactoring much easier and help you fit it into your sprint planning. In this section, we will have a look at a couple of examples of code refactors in the BookSwap application that will allow us to use all the techniques we have explored in this chapter.

Automated refactoring

One of the biggest strengths of Go is its tooling, and IDE support is no exception to this:

- The Google Go team maintains an extension for Go development in Visual Studio Code (`https://code.visualstudio.com/docs/languages/go`)

- The `vim-go` plugin (`https://github.com/fatih/vim-go`) is a popular open source plugin maintained by the Go community

- The team at JetBrains has created GoLand (`https://www.jetbrains.com/go/`), which is a dedicated product for Go development

All of these IDEs provide us with support for looking up references and usages of a given type and for renaming symbols across the entire call stack. This can take a lot of the boring grunt work of simple refactoring, but you will still have to make quite a few changes yourself.

Let's consider the refactoring involved in renaming `BookService` to `BookRepository`. We might want to change this name, as we added functionality related to the database of the `BookSwap` application in *Chapter 6, End-to-End Testing the BookSwap Web Application.*

First, we will rename the struct with our IDE's rename symbol functionality:

```
// BookRepository contains all the functionality and
// dependencies for managing books.
type BookRepository struct {
  DB *gorm.DB
  ps PostingService
}
```

This will update all the direct references to the old `BookService` in all implementation and test code across the entire application. This saves us from fixing a lot of compilation errors manually.

Next, we need to ensure all methods relating to this struct are correctly named. The `NewBookService` initialization function will need to be renamed as well:

```
// NewBookRepository initialises a BookService given its
// dependencies.
func NewBookRepository(db *gorm.DB, ps PostingService)
*BookRepository {
  return &BookRepository{
    DB: db,
    ps: ps,
```

```
    }
  }
```

The renamed function makes it clear that it is responsible for creating `BookRepository` given its dependencies.

We will need to review any test code for test signatures that relate to the old name as well. As we want to name tests after the functionality they verify and not the types they verify, we will not need to change any test names for this rename.

Finally, the filenames that contain and test these definitions will need to be changed to match:

- The `book_service.go` file becomes `book_repository.go`, making the naming of the file and the code it contains consistent

- The `book_service_test.go` file becomes `book_repository_test.go`, ensuring that the test code and implementation stay grouped together

That's all the work we need to do for renaming a service in our `BookSwap` application. This simple code refactor did not require any test changes, but it did demonstrate the process that you will have to undertake in refactoring Go code and how you can rely on your IDE to take care of some of the more laborious parts.

Validating refactored code

While renaming symbols is straightforward, a far more common change that you will find yourself having to make will be to change a method's signature. Let us see refactoring for a method signature change of the `Get` method of `BookRepository`, which currently has this signature:

```go
// Get returns a given book or error if none exists.
func (bs *BookRepository) Get(id string) (*Book, error)
```

This method takes in an ID, fetches `Book` from the database, or returns an error in the case that the book is not found. This is a common signature for this functionality.

We will change this method to take in `*Book` and return only an error. This will mean that the book fetched from the database will be populated on the book parameter and return an error if is not found. The new signature of this method will be as follows:

```go
// Get populates a given book or returns error if none
// exists.
func (bs *BookRepository) Get(b *Book) error
```

With the new signature in place, it's time to change our test code accordingly. The assertions in `TestGet`
`Book` of the `book_repository_test.go` file get changed to make use of the new signature:

```
for name, tc := range tests {
  t.Run(name, func(t *testing.T) {
    var b db.Book
    b.ID = tc.id
    err := bs.Get(&b)
    if tc.wantErr != nil {
      assert.Equal(t, tc.wantErr, err)
      assert.Nil(t, b)
      return
    }
    assert.Nil(t, err)
    assert.Equal(t, tc.want, b)
  })
}
```

We change the test code to make it compile and adjust to the new signature of the function. During
refactoring, tests should be changed as little as possible to ensure that the refactored code has not
caused any regressions.

At this point, the test will be compiling, but we have not completely implemented the code for the
new signature. It's time to turn our attention to the implementation of this new method:

```
func (bs *BookRepository) Get(b *Book) error {
  if r := bs.DB.Where("id = ?",
    b.ID).First(&b); r.Error != nil {
      return r.Error
  }
  return nil
}
```

The code is adjusted to read the ID of the b *Book parameter, and the database populates its results
to the same passed-in parameter. Then, we return an error or `nil` according to whether the item is
found or not.

Any other calling code will need to be adjusted in the same way as we have adjusted our test code. The compiler will let you know if you have missed any code that needs to be refactored. For example, if we add a second parameter to the `Get` method of `BookRepository` and forget to change it in its test, the compiler will let us know that the expected method signature is not defined when the test is run:

```
$ go test -run TestGetBook ./chapter07/db
./book_repository_test.go:34:19: not enough arguments in call
to bs.Get
    have (*db.Book)
    want (*db.Book, string)
FAIL    github.com/PacktPublishing/Test-Driven-Development-in-
Go/chapter07/db [build failed]
```

Remember, your test code is the first external consumer of your package's API, so any changes to the implementation code will affect your tests first. Note that this non-compiling state of the code has not been committed on our repository, so your test output will not match the preceding snippet.

Error verification

In *Chapter 4*, *Building Efficient Test Suites*, we briefly discussed Go's approach of explicit error handling. We learned that errors are typically returned last in a list of multiple return values. So far, we have been using Go's inbuilt `error` type and representative error messages to indicate to the user when something has gone wrong. Let us now take a closer look at how error verification works in Go.

We have created errors in two ways so far. The simplest way is using the `errors.New` function. It creates an error with a given message:

```
err := errors.New("Something is wrong!")
```

This function takes in an error message as a parameter and returns the `error` interface type. In order to get our error message back, we invoke the `Error` method on the function:

```
msg := err.Error()
```

This method returns the message as a string type.

Writing a test to compare the incoming and outgoing error messages is trivial for this example:

```
func TestErrorsVerification(t *testing.T){
    t.Run("simple custom error", func(t *testing.T) {
        wantMsg := "Something went wrong!"
        err := errors.New(wantMsg)
```

```
      gotMsg := err.Error()
      assert.Equal(t, wantMsg, gotMsg)
   })
}
```

As we have full control of the entire error message, we can easily assert that the values are equal. However, what if the error message construction is part of the function? It is common practice in Go to construct representative error messages that include inputs and other call parameters. These kinds of errors are typically constructed using the fmt.Errorf function:

```
func checkOdd(input int) error {
   if input%2 == 0 {
      return fmt.Errorf("Input %d cannot be even.", input)
   }
   return nil
}
```

The fmt.Errorf function formats the error the same as the rest of the formatting functions in the fmt package, but returns an error type with a well-formatted message. Asserting on this error message is slightly more complicated.

The first option is to reconstruct the error message in the test code:

```
func TestErrorsVerification(t *testing.T) {
   t.Run("formatted custom error", func(t *testing.T) {
      input := 4
      wantMsg := fmt.Sprintf("Input %d cannot be even.",
         input)
      err := checkOdd(input)
      gotMsg := err.Error()
      assert.Equal(t, wantMsg, gotMsg)
   })
}
```

The test code makes use of the fmt.Sprintf function to format the expected error message using the same format from the call to the fmt.Errorf function in the checkOdd implementation function.

This first option has three disadvantages:

- The test code has to repeat the implementation code simply for the purposes of verification. This can get complex if the error message requires significant setup.

- The test code is now tightly coupled to the implementation code. Changing the error formatting logic in the implementation code now requires the same change in the test code.

- There is no way to ensure that the formatting is replicated in the exact same way across test scenarios. This is likely to be a problem in large code bases that are maintained by large engineering teams.

The second option is to relax our error verification so that we no longer need to completely recreate the error message:

```go
func TestErrorsVerification(t *testing.T) {
    t.Run("formatted custom error", func(t *testing.T) {
        input := 4
        err := checkOdd(input)
        gotMsg := err.Error()
        assert.Contains(t, gotMsg, fmt.Sprint(input))
        assert.Contains(t, gotMsg, "even")
    })
}
```

The assert.Contains function is used to verify that the error message contains some substrings, which we can be relatively sure will not change in the implementation code. This option has removed the need for full implementation of the error message formatting, simplifying our test code.

However, this second option also has quite a few disadvantages:

- The error message assertion is not fully verified. For example, the implementation could be producing completely nonsense messages and our test will pass as long as the strings verified are contained within it. The test would no longer be able to assert on the full functionality of the implementation code. Other types of testing, such as integration or end-to-end tests, may verify this. However, we will explore how to include error assertions in unit tests.

- Even though it has been reduced, the test code still has some leaked implementation knowledge and a hardcoded part of the error message. Therefore, the test is still brittle and tightly coupled to its implementation counterpart.

- There is still no way to ensure that the assertions are performed in the same way across tests. In fact, because the expected string is no longer constructed during the Arrange part of the test, it can be even more difficult to find the hardcoded strings until the test suite points out the failures.

As discussed, both of our two immediate options have significant disadvantages. However, as we remember from *Chapter 4*, *Building Efficient Test Suites*, the `error` type is a simple interface with only a single method:

```go
type error interface {
  Error() string
}
```

We can easily implement our own custom error types by implementing this simple function. As we have seen on multiple occasions, the power of interfaces shines in many aspects of the Go programming language, and error handling is one of them as well.

Custom error types

The third option for error handling is to create our own custom error types, which will allow us to add more information to the error type than simply formatting a string multiple times. This will give us flexibility in both implementation and test code.

First, we will define a simple `evenNumberError` type:

```go
type evenNumberError struct {
  input int
}
```

This type has a field for the input of our `checkOdd` function. This will allow the test to have access to the input value without having to check the returned error message, which was necessary for the first and second options presented previously.

Next, we need to add a method to this new type to ensure that it satisfies the `error` interface:

```go
func (e *evenNumberError) Error() string {
    return fmt.Sprintf("Input %d cannot be even.", e.input)
}
```

This method has `evenNumberError` as a receiver and the same signature as the `error` interface. Inside the method, we use the same format and the `fmt.Sprintf` function, together with the `input` field of the receiver.

The implementation function can be changed to use this new `error` type:

```go
func checkOdd(input int) error {
  if input%2 == 0 {
    return &evenNumberError{
```

```
        input: input,
      }
    }
    return nil
}
```

With the error formatting wrapped inside the `evenNumberError` type, the `return` statement of this function simply consists of creating a new instance of this type and returning a pointer to it. We pass the parameter from the `checkOdd` function to its initialization.

> **Always return the error interface**
>
> One last thing to note is that the `checkOdd` function still returns the `error` interface. Therefore, the calling code does not need to have any knowledge of the custom error types created in this package. When working with custom error types, you should always follow this pattern as well.

The test code is much simplified with this new custom `error` type in place:

```
func TestErrorsVerification(t *testing.T) {
    t.Run("custom error type", func(t *testing.T) {
        input := 4
        wantErr := &evenNumberError{
            input: input,
        }
        err := checkOdd(input)
        var gotErr *evenNumberError
        require.True(t, errors.As(err, &gotErr))
        assert.Equal(t, wantErr, gotErr)
    })
}
```

The implementation of the error verification in this example demonstrates how to run verifications on custom `error` types:

1. We create an instance of the `evenNumberError` type with the `input` field. This is much simpler than having to create an expected error message.

2. After calling the `checkOdd` function, we need to convert the built-in `error` value to the custom `error` type. This is done by using the `errors.As` function, which returns `true` if the conversion has been successful.

3. We use the `require.True` function to ensure that the test fails if the conversion fails.

4. Finally, we use the `assert.Equal` function to ensure that the actual error is as expected.

The implementation of the test is much simpler, and it is no longer tightly coupled to error formatting inside the function under test. This approach does have a slight disadvantage in that it creates a new custom type, but working with custom `error` types streamlines the implementation code by providing a single, unified, way to format errors.

We run the test as usual to see our error verification in action:

```
$ go test -run TestErrorsVerification ./chapter07/errors -v
=== RUN    TestErrorsVerification
=== RUN    TestErrorsVerification/simple_custom_error
=== RUN    TestErrorsVerification/formatted_custom_error
=== RUN    TestErrorsVerification/contains_custom_error
=== RUN    TestErrorsVerification/custom_error_type
--- PASS: TestErrorsVerification (0.00s)
   --- PASS: TestErrorsVerification/simple_custom_error (0.00s)
   --- PASS: TestErrorsVerification/formatted_custom_error
(0.00s)
   --- PASS: TestErrorsVerification/contains_custom_error
(0.00s)
   --- PASS: TestErrorsVerification/custom_error_type (0.00s)
PASS
ok        github.com/PacktPublishing/Test-Driven-Development-in-
Go/chapter07/errors        0.122s
```

Each test case runs in its subtest, as can be seen from the organized output.

Another advantage of using custom `error` types is that they allow a function to return multiple types of errors, which can provide context to callers of a given package or service. For now, we should remember that they have the advantage of streamlining our test code while providing precise error verification possibilities.

Splitting up the monolith

The final aspect of refactoring that we will discuss is moving from a monolith to a microservice architecture. While there are examples of large companies that successfully operate using a monolithic architecture, the consensus in the technical community is that a microservice architecture is easier to scale and maintain, particularly when working across multiple teams. It is therefore important to discuss some of the basics of how and when to split up a service during refactoring.

> **What is a monolithic application?**
>
> A **monolithic application** is a single application that is built and released in one unit. The term is typically used to refer to a large application, with many different responsibilities that serve many different user journeys.

Figure 7.5 depicts some of the advantages and disadvantages of monolithic applications:

Pros	Cons
+ Easy to deploy	- Harder to scale
+ Lower cognitive burden	- Slower development speed
+ Simple to test and debug	- Errors can cause full outage

Figure 7.5 – Pros and cons of monolithic applications

The list of pros and cons of monolithic applications spans from deployment to resilience:

- As monoliths are built and deployed in one running application, developers will find them **easy to deploy**. However, as all the components are deployed together, they also cannot be scaled individually, making the application **harder to scale**. This can be a significant bottleneck to how many requests the application can handle, affecting a business's revenue. Scaling the monolith as a whole can also be unduly expensive, as all resources are scaled together.

- Monolithic applications have a lower cognitive burden because all of their code lives in one single, searchable code base. At the beginning of the project, developers have an easier time developing in one single code base. However, as the project progresses and the team grows, the code base becomes constrained by its initial architecture, design, and technology choices. Often, this leads to **slower development speed**.

- As they only have to start up one application, engineers find monolithic applications **simple to test and debug**. However, as all modules are hosted together, **errors can cause full outages** of the application. This can again significantly affect a business's revenue.

Many organizations start out with a monolithic application when they have a small code base and engineering team. Then, as their team and application functionality grow, they struggle to scale and maintain their monoliths. A new approach to building applications was required to mitigate the disadvantages that come with developing code in a monolith.

> **What is a microservice architecture?**
>
> A **microservice architecture** is a system design method that relies on independently built and released services, known as **microservices**. These independent units have their own responsibilities and self-contained resources for accomplishing their goals.

Figure 7.6 depicts some of the advantages and disadvantages of microservice architectures:

Pros	Cons
+ Flexible scaling	- Higher infrastructure costs
+ Easier to maintain	- Higher organizational overhead
+ Smaller scope of tests	- More difficult to debug

Figure 7.6 – Pros and cons of microservice architectures

The pros of microservice architectures have addressed many of the cons of monolithic applications:

- As each microservice is independently deployed, this type of architecture offers **flexible scaling**. This allows us to scale one part of the application, according to which user journeys are most popular. However, as each service has its own dedicated set of resources, microservice architectures can incur **higher infrastructure costs**.

- The smaller code bases of the microservices are **easier to maintain,** especially when it comes to refactoring. They can also make their own technical choices, offering engineers the opportunity to choose the best tool for the goals of each microservice. However, the separation between the different code bases does require **higher organizational overhead** in order to ensure that the individual units function together correctly and follow a unified set of engineering standards.

- As each microservice has its own well-defined functionality and responsibilities, they require a **smaller scope of tests** to ensure that they are working correctly. However, the integration points become the focal point of tests, making integration tests more important than ever. When a systemic error happens, each service also has its own set of logs, making this type of architecture **more difficult to debug**.

Figure 7.7 depicts the two system types:

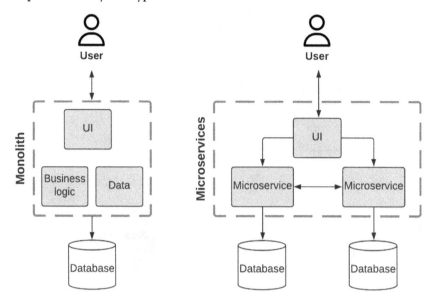

Figure 7.7 – Monolithic application versus microservice architecture

The monolithic application contains all of the components required in one single unit and relies on one database. The microservice architecture splits the monolithic application according to the functionality that they provide and these functionalities depend on each other to deliver the same user journeys as the monolithic application.

As we have seen, neither development approach is perfect. A monolithic application may work perfectly well when application traffic is predictable or the team has strict architecture design guidelines, but most organizations grow their teams and evolve their products. In these circumstances, organizations will usually opt for a microservice architecture. Therefore, it is absolutely essential for engineers to know how to develop and test microservice architectures.

Key refactoring considerations

Many organizations have undertaken the work to split up their monolithic applications into microservice architectures. The engineering community has discussed how to undertake this journey as painlessly and successfully as possible. We can highlight some key considerations for this type of refactoring.

Define boundaries

In order to be successful, microservices need to have their own domain and have well-defined application boundaries. Engineering teams can identify which parts of the monolith they should extract, either through analysis or generated dependency graphs.

Based on these, they can scope the following:

- The **functionality and models of the domain** that the microservice will be responsible for. For example, in an e-commerce application, we might identify a service that is responsible for placing and managing items in the shopping cart.

- The **upstream dependencies** that the microservice will require. For example, the previously identified shopping cart service will have a dependency on the inventory service, which will tell it how much items cost and whether they are in stock.

- The **downstream dependencies** and data storage solution that the microservice will require. For example, the shopping cart service will save its data to an in-memory data structure store, such as Redis, and have a downstream dependency to a checkout service once the user decides to purchase their cart.

The result of the boundary identification exercise could be a high-level design of the infrastructure requirements of the microservice, as well as an overview of the API that the microservice will expose to other parts of the system.

Loose coupling

One of the advantages of a microservice architecture is that it allows us to build services that are **loosely coupled**, from both a development and deployment perspective. However, this advantage can be easily lost if the teams don't isolate inter-service dependencies.

> **What is a loose coupling for microservices?**
>
> In microservice architectures, services are loosely coupled if changes to one service's design or implementation will not cause changes in other services that it depends on or that depend on it.

Loosely coupled microservices should follow the following rules of thumb:

- **Have separate data stores**: When microservices share a single database, they also share a **single point of failure**. An outage on the shared database will cause all of the services that depend on it to fail as well.

- **Use asynchronous communication**: Use asynchronous communication patterns such as queues and event buses to pass data to other services. This allows us to scale services independently and even batch messages when appropriate.

- **Implement fault tolerance**: Design your service with the assumption that its internal and external dependencies will fail or be slow to respond. A common way this is implemented in REST APIs is with the circuit breaker pattern, which will time out external calls when patterns of failure are detected in order to ensure that your microservice can continue to function, using default values if one of its dependencies is no longer available.

- **Make backward-compatible changes**: Whenever possible, microservice APIs should make backward-compatible changes that will not force other services to implement changes to their payloads at the same time.

- **Implement request tracing and service monitoring**: In the microservice world, it is important to understand how requests flow through your system. This is important for detecting errors in your system and identifying services with insufficient infrastructure resources.

Following these rules during microservice design and implementation will help you take advantage of the true power of microservices. In *Chapter 8, Testing Microservice Architectures*, we will discuss these principles further and discuss the challenges of testing microservices and how we need to adjust our testing strategy to fit the microservice world.

> **How will the microservice behave without its dependencies?**
>
> Once we have identified the dependencies of a microservice, we should remember to design and test for the behavior of the microservice in the case that its dependencies encounter an outage. This important part of microservice design, known as **graceful degradation**, should not be overlooked.

Summary

In this chapter, we tackled the important topic of code refactoring, which is a crucial and unavoidable part of extending and maintaining healthy code bases. We started by learning some common code refactoring techniques and discussed the true cost of technical debt. Then, we revisited the power of interfaces, which make it easy to change dependencies and allow us to use the compiler as a guide during refactoring.

Then, we considered the test changes that we have to make to our tests to ensure that they continue to verify behaviors during two common refactorings: renaming structs and changing method signatures. Expanding upon our previous knowledge of error handling and verification, we learned how to create custom error types and more easily verify error messages.

Finally, we learned some of the reasons why organizations move from monolithic applications to microservice architectures, and explored some rules of thumb that allow us to create loosely coupled microservices.

In *Chapter 8, Testing Microservice Architectures*, we will expand on all the concepts we have learned so far and learn what considerations should be made when testing microservice architectures. We will apply and demonstrate these concepts on our demo application, the BookSwap application.

Questions

1. What is the difference between code redesign and code refactoring?

2. Describe the working process of code refactoring.

3. What is technical debt?

4. What is a monolithic application? What is a microservice architecture?

Further reading

- *Clean Architecture: A Craftsman's Guide to Software Structure and Design*, Robert C. Martin, published by Addison-Wesley

- *Refactoring: Improving the Design of Existing Code*, Martin Fowler, published by Addison-Wesley Professional

- *Building Microservices Second Edition: Designing Fine-Grained Systems*, Sam Newman, published by O'Reilly

- *Monolith to Microservices: Evolutionary Patterns to Transform Your Monolith*, Sam Newman, published by O'Reilly

8

Testing Microservice Architectures

The topics we have covered so far have gone beyond the scope of how to write tests. We have looked at a wide range of software design and development concerns, including containerization with Docker and database integration with PostgreSQL. This highlights the fact that writing good tests requires a thorough understanding of the architecture and technical dependencies of the application under test.

Alongside these software development concepts, we discussed the evolution of code in *Chapter 7, Refactoring in Go*. We learned some common refactoring techniques, and we compared monolithic applications with microservice architectures, which is a common evolution of Go web applications as they grow and become more mature.

We will continue our exploration of microservice architectures and refactoring from the previous chapter. As microservices are often owned and developed by different software teams, they are often changed without any central oversight. In this fast-paced world of changing requirements and implementations, ensuring that the API integration points in our system are still functioning correctly is one of the biggest challenges to overcome. Another key concern to consider is error detection in the system – when something goes wrong, how do you isolate the malfunctioning service in a large map of dependencies?

This chapter is dedicated to discussing the testing of microservice architectures, demonstrated on the monolithic BookSwap web application introduced in previous chapters. We will have a closer look at the implementation of non-functional tests, which was briefly discussed in previous chapters. Then, we will learn the new concepts of contract testing and how we can leverage Pact for the implementation of contracts on microservice architectures. Finally, we will discuss how we can split up the monolithic BookSwap web application that we have built so far. Using the concepts and challenges we will have learned, we will discuss some best practices for running microservices in production.

In this chapter, we will cover the following topics:

- The implementation of non-functional testing
- Challenges of testing microservice architectures
- Getting started with contract testing with Pact
- Splitting up the `BookSwap` monolith we have built so far
- Best practices for running microservice architectures in production

Technical requirements

You will need to have **Go version 1.19** or later installed to run the code samples in this chapter. The installation process is described in the official Go documentation at `https://go.dev/doc/install`.

The code examples included in this book are publicly available at `https://github.com/PacktPublishing/Test-Driven-Development-in-Go/chapter08`.

Functional and non-functional testing

We briefly touched upon the topic of non-functional testing in *Chapter 1, Getting to Grips with Test-Driven Development*. Up until now, we have tabled this important type of testing and focused on verifying the various functional aspects, while exploring the popular testing libraries of `testify`, `ginkgo`, and `GoDog`. Let's now explore how to implement a few of the most important non-functional tests.

Figure 8.1 depicts the main types of non-functional tests:

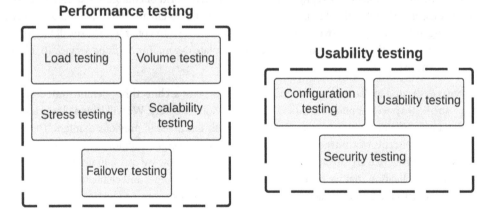

Figure 8.1 – Types of performance and correctness non-functional testing

The types of tests are divided between **performance tests** and **usability tests**. They verify the following aspects of our systems:

1. **Load testing** simulates user demand on our system. These tests simulate expected demand and overload conditions to identify bottlenecks or performance issues.

2. **Stress testing** simulates user demand under extreme conditions on our system. These tests are used to identify the scalability limit of our system and verify that it handles errors gracefully when components become overloaded.

3. **Volume testing** simulates large volumes of data coming into our system. This is similar to stress testing but with a few tests, each involving relatively large amounts of data, instead of many tests involving smaller amounts of data simulating user demand. These tests are used to identify the data limits that our system can process, which is particularly useful for services with a database/persistent storage solution.

4. **Scalability testing** verifies our system's ability to scale its components when subjected to sudden load. The load can be applied gradually, or it can be applied suddenly, which is known as a **spike test**.

5. **Failover testing** verifies our system's ability to recover after a failure. This type of negative testing is a useful simulation for how quickly the system can recover following incidents.

6. **Configuration testing** verifies our system's behavior with different types of settings. They can be user-controlled settings or system settings. The system setup can change the expected behavior of the system, as well as its performance.

7. **Usability testing** verifies how intuitive the user-facing functionality is to use. The focus of this type of testing varies according to the functionality that the system exposes, but it typically covers the following:

   I.   How intuitive the system is to use for new users

   II.   How easily users can perform their tasks

   III. Whether error messages are well formulated and guide the user

8. **Security testing** verifies whether security practices have been followed during the development process. The system under test should have correct authentication, authorization, and data integrity features.

As we have seen, non-functional tests are extremely important for ensuring that our systems are functioning correctly under a wide variety of conditions. No testing strategy is complete without covering some of these important types of tests.

> **Non-functional tests verify crucial aspects of our system**
>
> These tests verify the performance and usability of the system under test, including how well the system scales and recovers from outages. These types of tests might be performed by different development teams, as they might require skills from outside the engineering team to implement them.

Performance testing in Go

While we have already established that non-functional testing covers important aspects, performance testing becomes even more important when moving from monolithic applications to microservice architectures. In the microservice world, the user journey varies and is processed by independent system components, which can lead to a less cohesive view of system behavior.

Figure 8.2 depicts the key questions that performance testing answers:

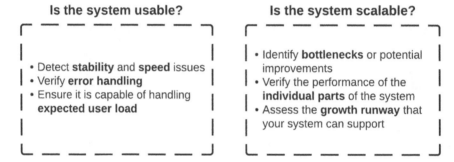

Figure 8.2 – Key questions that performance testing answers

The two important questions that performance testing answers relate to system usability and scalability. Let's look at what each question means.

Is the system usable?

Usability is more than achieving correct functionality, as a slow-functioning system will eventually impact user satisfaction negatively. Performance testing is useful for assessing the following:

- **Stability**: No intermittent failures should occur, causing retries and negative user experiences.
- **Speed**: User requests should be kept within acceptable levels set according to business requirements, or the system is scaled appropriately.
- **Error handling**: Errors should be handled gracefully, without sudden crashes, and well-formulated messages should be returned across a variety of scenarios.
- **User load**: The system should be able to handle the expected user loads without causing unexpected CPU or user memory spikes.

Is the system scalable?

Business and system requirements evolve with time. A scalable system should be able to grow according to the expected future needs of the business. Performance testing is useful for assessing the following:

- **Bottlenecks**: Monitoring a variety of metrics allows us to identify which services in our system are not scalable, and should be refactored.

- **Individual parts**: It is important to understand the expected response time for each microservice, as well as an estimate for the entire system. This can help us map costs for each user operation on our system.

- **Growth runway**: Performance testing allows us to establish how much more user and volume growth the system can sustain in its current form.

When used correctly, performance testing will ensure that each microservice is able to handle the current load of the system and that they are able to work together to serve user journeys correctly.

> **The "little and often" approach**
>
> Performance testing is often added as part of the code build pipelines so that development teams get immediate feedback on performance with each commit. Similarly to refactoring, performance improvements are best done little and often. Monitoring performance with each commit will make it easier to see any trends and quickly fix new issues.

Performance testing is all about quantifying and comparing the behavior of our system and its microservices. How do we go about achieving this quantification? This is commonly achieved by gathering a few important metrics:

- **Response time**: The time it takes between a user's request and the response from the system to arrive back to the user. Often, the **average** and **peak** values of the response time are measured, giving an indication of the worst case alongside the average case.

- **Error rate**: The percentage of error cases in the total number of requests processed by the system. In RESTful APIs, the error responses are easily identified by HTTP status codes.

- **CPU and memory usage**: The percentage of CPU and memory that the microservice is using on its host. These indicators will show whether the system is correctly scaled.

- **Concurrent users**: The number of users that are requesting a given resource at the same time. This can make it easy to identify any spikes for a particular endpoint of the user path.

- **Data throughput**: The amount of data processed by the system. This can indicate whether user requests are increasing over time or whether any large files are flowing into the system and affecting performance.

The system under test should have monitoring and alerting for these metrics in place before we write any performance tests. Furthermore, we should establish what the failure criteria for our performance tests will be according to the needs of our system.

While you should always establish your threshold values together with key stakeholders, we can make some general recommendations based on experience and industry practice:

- The average response time should generally be under 500 milliseconds, while the peak response time should be under 1 second
- Error rates should generally be under 5%
- CPU and memory usage should generally stay under 70%, allowing the system to handle any spikes that may come up
- Concurrent users and data throughput do not have any failure thresholds, but should be monitored for spikes and anomalies

Now that we understand the importance of performance tests and how to quantify and compare their results, we can turn our attention to their implementation. We can implement them with Go's standard `testing` framework or with popular third-party libraries.

Implementing performance tests

In *Chapter 2, Unit Testing Essentials*, we learned how to write and execute benchmarks with Go's standard testing library, which are special tests used to verify the performance of our code. We also learned how to export test coverage metrics from Go's test runner.

We can use benchmarks to write performance tests for our endpoints. For example, we can easily write a benchmark for the GET / root endpoint of our BookSwap application:

```
func BenchmarkGetIndex(b *testing.B) {
    endpoint := getTestEndpoint(b)
    for x := 0; x < b.N; x++ {
        bks, err := http.Get(*endpoint)
        assert.Nil(b, err)
        assert.NotNil(b, bks)
    }
}
```

We create a new benchmark according to the expected signature, taking in a single `*testing.B` parameter and named with the `Benchmark` prefix. Then, we make use of the standard `http` library to invoke the GET operation on the defined endpoint, which is returned by the `getTestEndpoint` helper function. Just as in previous chapters, this function constructs the endpoint based on the provided environment variables. If you want to run with the default values, set the BOOKSWAP_BASE_URL environment variable to http://localhost and the BOOKSWAP_PORT environment variable to 3000 to your terminal session.

We save this test in the chapter08/performance/books_index_test.go file. With our simple test written, we need to make sure that the BookSwap application is up and running. We can easily run it using the docker compose -f docker-compose.book-swap.chapter08.yml up --build command. As mentioned in previous chapters, remember to set the BOOKSWAP_PORT environment variable before running. If you are running with default configuration, then you can use 3000 for its value.

Next, we need to run the benchmark. The go test command provides support for profiling benchmarks in a similar way to how we extracted code coverage details, in *Chapter 2, Unit Testing Essentials*. The runtime/pprof package provides the following predefined profiling options:

- cpu shows us where our program is using CPU cycles
- heap shows us where our program is making memory allocations
- threadcreate shows us where the program is requiring new threads
- goroutine shows us stack traces of all the program's goroutines
- block shows us where goroutines are waiting on locking primitives
- mutex reports lock contention

We will explore the concurrency aspects of threads, goroutines, and mutexes in *Chapter 9, Challenges of Testing Concurrent Code*. For now, we will focus on CPU profiling.

We run our newly written benchmark with two profiling options, which will allow us to extract the CPU profile:

```
$ go test -bench BenchmarkGetIndex -cpuprofile cpu-books.out ./
chapter08/performance
```

The benchmark runner outputs the same results we saw in our introductory chapter:

```
pkg: github.com/PacktPublishing/Test-Driven-Development-in-Go/
chapter08/performance
BenchmarkGetIndex-8              1556                 796124 ns/op
PASS
ok       github.com/PacktPublishing/Test-Driven-Development-
in-Go/chapter08/performance  2.600s
```

As our index endpoint is quite simple, the benchmark is executed 1,556 times and the total running time is 2.6 seconds. This command runs the benchmark and instructs the test runner to save the

CPU profile to the `cpu-books.out` file, saved in the current running directory. The details of the test run are saved in the `performance.test` file, which is named after the package that the test is declared in.

We can view the file using the `pprof` command tool, which comes installed with the Go toolchain:

```
$ go tool pprof performance.test cpu-books.out
```

This opens up an interactive command that will allow us to get some insights into the measured CPU time. The command will give a text output of the top profile results, while `web` will create a visual representation of the same results. Running `top5` on the CPU profile of our benchmark presents the following five results:

```
(pprof) top5
Showing nodes accounting for 550ms, 80.88% of 680ms total
Showing top 5 nodes out of 91
   flat  flat%   sum%    cum   cum%
  180ms 26.47% 26.47%  180ms 26.47% runtime.pthread_cond_signal
  120ms 17.65% 44.12%  120ms 17.65% runtime.kevent
  100ms 14.71% 58.82%  100ms 14.71% runtime.cgocall
   80ms 11.76% 70.59%  120ms 17.65% runtime.pthread_cond_wait
   70ms 10.29% 80.88%   80ms 11.76% syscall.syscall
```

These top results count for more than 80% of the running time, but they seem to be related only to the running and scheduling of the benchmark test itself. As the benchmark is scheduled and runs thousands of times, we can expect that the test runner will need to make use of quite a few goroutines and threads to execute the test. However, this is not very useful output for gaining an understanding of the operation of our `BookSwap` web application. We cannot profile our web application from the benchmark test since the web application is running in a whole other process, separate from the benchmark.

In order to gain insights into the CPU usage of our `BookSwap` application, we will need to integrate the `pprof` tool into our web application. This is easy to do by allowing `pprof` to register itself alongside our other handlers in `handlers/config.go`:

```go
func ConfigureServer(handler *Handler) *mux.Router {
    router := mux.NewRouter().StrictSlash(true)
    // other handler functions
    if os.Getenv("DEBUG") != "" {
        router.PathPrefix("/debug/pprof/").
            Handler(http.DefaultServeMux)
```

```
}
    return router
}
```

pprof will now be able to serve all the paths configured with the debug/pprof prefix if the DEBUG environment variable is set when the application is started. We can easily set it by adding the line DEBUG=true to the docker.env. We can then rerun the application in debug mode using the docker compose -f docker-compose.book-swap.chapter08.yml up --build command. This allows us to selectively expose this endpoint in particular environments. We are now ready to profile our web application. We rerun our benchmark, which will take around 3 seconds to run. We can then download the results to a local file in the same way that we exported the results of the benchmark profile:

```
$ curl --output book-swap-app "http://localhost$BOOKSWAP_PORT/
debug/pprof/profile?seconds=10"
```

As the application is running locally in this example, the URL is localhost:$BOOKSWAP_PORT, but we would change it for other environments and configurations. This command downloads the profiling data from the past 10 seconds and saves it to a local file. We can then view the exported results in the same way as before:

```
$ go tool pprof book-swap-app
```

This command opens up the same interactive screen as before, but we will now opt to see the visual representation of the CPU using the web command. This will launch a window in your default browser with a graph of the method calls that have been profiled.

Graph visualization

Go's profiling tool, pprof, relies on an external dependency for graph visualization. This dependency, named graphviz, is not written in Go and is therefore not automatically installed with the Go toolchain. You should follow the official documentation (https://graphviz. org/download/) to install it for your operating system.

Figure 8.3 presents the CPU profile usage in a visual representation, as was measured during the benchmarking of the index endpoint:

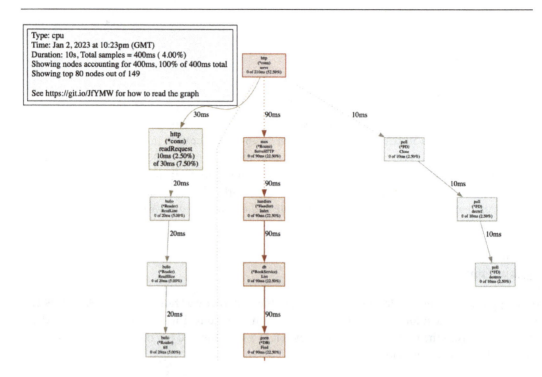

Figure 8.3 – A visual representation of the BookSwap CPU profile

As we can from the CPU profile, the BookSwap application spends most resources serving HTTP connections with net/http and interacting with the database layer using the GORM library. This is indicated by the percentages and size of the boxes corresponding to each operation. The visual representation of the call stack gives us a good indication of where we are spending our resources. We explored the database aspects of the BookSwap application in *Chapter 6, End-to-End Testing the BookSwap Web Application*. If we want to improve the performance of the application, we can use the information presented in the profile to identify areas of the call stack that need to be improved.

While benchmarking allows us to create simple tests and simulate a variety of load-testing scenarios, it can be quite verbose to define testing scenarios across many different microservices. There are two popular open source libraries that are often used for performance testing:

- **JMeter** (https://jmeter.apache.org/) is an open source Java testing tool maintained by Apache. Test plans are recorded using a simple UI, removing the need to write boilerplate code with Go's testing package. Different types of load can be configured. JMeter also has the capability of generating result graphs and dashboards once the tests are run.

- **K6** (`https://k6.io/`) is an open source Go project maintained by Grafana. Test plans are written in a scripting language similar to JavaScript, reducing a lot of the code needed to write test scenarios. K6 offers different types of load configurations and also has the capability of outputting test results to dashboards.

- **Gatling** (`https://gatling.io/open-source/`) is an open source Scala load testing tool maintained by Gatling Corp. Similarly to K6, tests are written in a Domain-Specific Language, but it is based on Scala. This library provides load testing and insights on dashboards.

Regardless of which performance testing implementation option you choose, you can profile your application and supplement the data and graphs that it supplies. We will not be exploring these third-party tools in this book, as we have used Go's in-built benchmarking capabilities to write our performance tests.

Go profiling is a very powerful tool with many more capabilities than what we have explored here. You can read more about Go's diagnostics capabilities in the official documentation (`https://go.dev/doc/diagnostics`).

> **Profiling tests and applications**
>
> While we did not directly use the profiling information of the benchmark we ran, profiling tests can be a useful way to investigate costly or slow-running tests. Therefore, knowing how to export and read profiling information is useful for both development and test writing.

Contract testing

As discussed in *Chapter 7, Refactoring in Go*, microservice architectures have many advantages over monolithic applications: the ability to scale system components independently, smaller code bases that are easier to maintain, and a system that is less prone to outages. However, the development and testing of working processes change when organizations adopt microservice architectures. This also brings challenges, alongside the vast benefits of microservice architectures.

Figure 8.4 depicts the three types of complexity that microservice architectures bring:

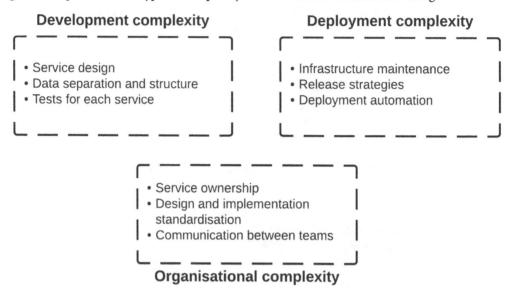

Figure 8.4 – The complexities of microservice architectures

Microservice architectures add complexity to every part of the development process:

1. **Development complexity**: The source code of each microservice is often contained in its own separate code base or repository. This leads to complexity in the development process due to the following components:

    I.   **Service design** must be consistent across multiple services. Each engineering team must design multiple services, as opposed to creating one monolithic application and then changing it.

    II.   Related to service design, the **data separation and structure** must be designed as well. Each microservice is in charge of saving its own data to persistent storage and sending the information to other services when they require it. If this is done without any design, services will need to pass data back and forth, increasing response times.

    III. Finally, the team will need to implement **tests for each service**. If the service exposes user-facing functionality, it will need to be tested at every level of the testing pyramid. This will increase the number of tests required for the system, even though they may be faster and test a smaller functionality scope.

2. **Deployment complexity**: Each microservice is its own self-contained running application. This leads to complexity in the deployment pipelines due to the following components:

  I.   The development teams have a higher burden of **infrastructure maintenance** due to the separation of each microservice and its dependencies. This can become even more complex when services require different kinds of dependencies or versions, as the system matures and the microservices are not updated at the same time.

  II. **Release strategies** become more complex when it comes to making changes, as dependencies inside the system become more complex. All updates to the data structure or API changes, including the services, are not directly user-facing, as they could cause outages elsewhere in the system.

  III. **Deployment automation** becomes a necessity in order to make it feasible for teams to easily build and release services. Testing must also be added to the release pipelines to minimize the risk of outages.

3. **Organizational complexity**: Teams are unblocked to develop and release multiple services at the same time. This leads to an increase in productivity, but also organizational challenges due to the following components:

  I.   Often, the number of microservices far outnumbers the number of engineering teams, and in some cases even engineers! Therefore, **service ownership** is extended to multiple services per team. This adds maintenance complexity to the teams, which they must manage alongside delivering new features.

  II. Teams must agree on a common way to structure and implement their services so that engineers can work across teams, as well as investigate services across the entire system. As such, the engineering organization will have to undertake some kind of design and implementation standardization process. This can be quite a difficult undertaking, as teams will have different requirements and/or preferences.

  III. Finally, **communication between teams** will need to effectively handle larger systemic changes in order to avoid outages. This can be difficult for teams that are growing rapidly.

The complexities introduced by microservice architectures can be mitigated with a solid testing strategy, which will flag any errors or breakages before they cause outages across the entire system. As discussed, the integration points between microservices must be tested, as teams will release changes to the services they own without any central oversight.

Figure 8.5 depicts how we might go about testing the integration between two microservices using the knowledge we have gained so far:

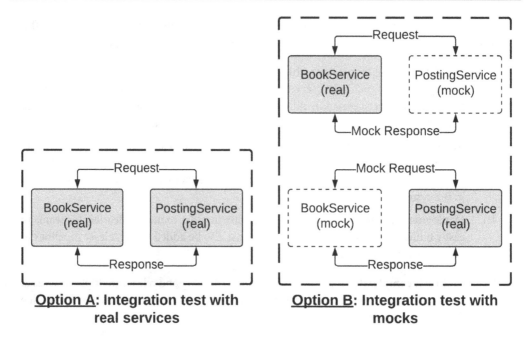

Option A: Integration test with real services   **Option B**: Integration test with mocks

Figure 8.5 – Testing the integration between two microservices

There are two options when it comes to testing the integration between two services:

- **Option A: Integration test with real services** involves writing an integration test between the real services in a testing environment. This approach allows us to verify that both services are functioning as expected and that their integration is successful. However, as the system grows, setting up each service and its dependencies becomes more complicated. Individual test runs will also slow down, as data and requests need to travel across multiple microservices or data stores.

- **Option B: Integration test with mocks** involves writing separate integration tests against mocks for the dependency. This approach allows us to reduce the scope of the test and ensure that each service is working as expected. However, as it tests each service in isolation, it does not actually verify that the services are working together as expected. If either service does not conform to its defined mock, then the test would pass even though we could be creating an outage. This is the same issue we identified with our mocks in *Chapter 3, Mocking and Assertion Frameworks*.

Neither of these two options is ideal because we would need to write robust tests that verify that our microservices are integrated well together to have the confidence to change microservices without central oversight. We will explore a third way of testing that can alleviate some of the downsides of each approach.

Fundamentals of contract testing

Due to the downsides of the existing solutions and the difficulties that come with testing microservice architectures, developers began using another type of testing practice. **Contract testing** offers a simpler way to ensure that microservices continue to integrate well. It is not a new concept, but it has gained traction because it is well suited for distributed architectures.

Developers write virtual contracts that define how two microservices should interact. This contract provides the source of truth and represents the expected values for test assertions. There are two sides to every contract:

- The **consumer** begins the interaction between the two microservices. The consumer issues the HTTP request or requests data from a message queue. In the example in *Figure 8.5*, BookService is the consumer as it sends the request.

- The **provider** completes the interaction between two microservices. The provider responds to the consumer's HTTP request or creates the message for the consumer to read. In the example in *Figure 8.5*, PostingService is the provider as it sends the response.

Based on this terminology, *Figure 8.6* demonstrates the procedure for writing and running contract tests:

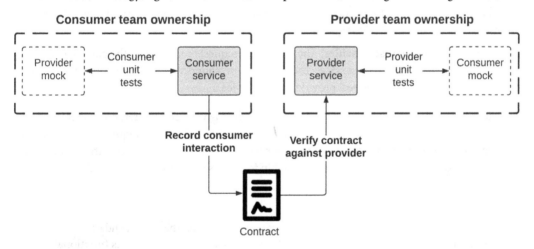

Figure 8.6 – Writing and running contract tests

The simple procedure consists of the following steps:

1. **Establish the consumer and provider**: We begin by identifying which services we want to test. In a microservice architecture, this isn't always straightforward. After all, there is no code coverage metric for distributed systems that we can rely on to see which microservice integrations haven't been tested.

2. **Identify the interaction(s) under test**: This step is equivalent to identifying which user journey we'd like to test or writing our feature test. This should include the HTTP method, the HTTP request body, and any URL parameters we might require. At this point, we should also establish what the expected response of the provider should be.

3. **Consumer unit tests**: As part of the development process, the team will write unit tests for the consumer service. This will be done against a provider mock that is under the **consumer team's ownership**.

4. **Provider unit tests**: In the same way as on the consumer service side, the team will write unit tests for the provider during the development process, we use a consumer mock that is under the **provider team's ownership**.

5. **Record consumer interaction**: Based on the identified parameters and interactions of the unit test, we can begin to formulate the contract between the consumer and provider. The consumer team captures the required interaction between services, which is made up of the consumer request(s) and the expected provider response.

6. **Contract**: The consumer request and provider response are recorded together in one file, known as the contract. It crosses team boundaries and is the source of truth for the two teams, allowing them to easily collaborate using a common language. As we mentioned previously, microservice architectures add organizational complexity so the contract can help teams communicate more effectively.

7. **Verify contract against provider**: The consumer requests recorded in the contract are run against the provider microservice. The expected provider response is verified against the response received from the real provider microservice.

A contract test is considered passed only when the contract is verified by interacting with the real services on both sides of the contract. However, unlike integration tests, which require one single team to have both the consumer and provider running for the test, contract testing allows this verification to be done in two steps, allowing the team ownership for each service to be maintained.

> **The consumer viewpoint**
>
> Contract testing is written starting with the consumer, which dictates the request and expectations. This helps us to ensure that the API is stable for the services that are using its functionality, encouraging stable APIs that do not promote breaking changes.

The contents of the contract file are the most important part of the process, and it is important that they does not contain any errors. The safest way to ensure that this does not happen is to use tools that help us generate them, as opposed to writing them manually. We will not attempt to implement contract testing manually, but instead, look at the process using one of the most popular tools.

Using Pact

Now that we understand the basic process of contract testing, we can have a look at some tools that facilitate the process by helping us generate contracts and run tests. Pact (`https://github.com/pact-foundation`) is a popular open source contract testing tool that allows us to easily write contract tests. It has been running since 2013, and it has quickly become the number-one choice for implementing contract tests.

Some of the main features of Pact are as follows:

- **Synchronous and asynchronous support**: Pact allows contract testing for HTTP endpoints, as well as asynchronous non-HTTP messaging systems. It supports a variety of technologies, such as Kafka, GraphQL, and publish-subscribe messaging patterns.

- **Libraries in over ten languages**: Pact offers support for a wide variety of languages for both frontend and backend technologies. The Pact Go library (`https://github.com/pact-foundation/pact-go`) provides us with the functionality required for testing our Go microservices.

- **Unit testing integration**: The consumer code base imports the Pact Go library and uses it to write unit tests. This allows developers to use the same workflow and techniques for contract tests as was used for writing unit tests.

- **Contract testing Domain-Specific Language (DSL)**: The Pact library gives projects a common DSL for writing contract tests. This allows developers to define interactions and expected responses in a uniform way.

- **Test playback and verification**: Based on the test specifications, Pact generates and records the test runs. Contract tests are called pacts, and they are replayed and verified against the provider service.

- **Broker service**: Pact provides a self-hosted broker solution that allows the easy sharing and verification of contracts and test results. This solution is suitable for production systems and integrating contract testing into the release pipelines.

This list of features is the reason why Pact has quickly become the contract testing tool of choice. We can easily implement the contract testing steps using the Pact Go library.

Pact provides a variety of command-line tools in an easy-to-install native binary that provides functionality for testing both synchronous and asynchronous message-based interactions:

- Find the newest version of the tools on the project release page (`https://github.com/pact-foundation/pact-ruby-standalone/releases`). This page will also contain installation instructions for your operating system.

- The Pact Go library supports Go modules and can be easily added to your projects with the usual command: `go get github.com/pact-foundation/pact-go`.

Adding the Pact tools to your system path

As detailed in the Pact setup instructions, remember to add the path to the pact/bin directory to your system path. The Go test runner will need to be able to call the Pact tools during the test running and verification.

The installation will install a few different tools that we can use during contract testing. You can explore them all on your own. Some of the most commonly used tools are as follows:

- pact-mock-service provides mocking and stubbing functionality. It can help us easily create mocks for our providers during contract testing.

- pact-broker provides functionality for starting up the previously mentioned broker service, which makes it easy to share contracts and verification results. It also allows you to deploy it independently, including using Docker.

- pact-provider-verifier provides verification of two versions of pacts, regardless of whether the values are coming from the Pact Broker or another source. The verifier is often added to the release pipelines, saving the development effort of implementing their own.

Once the tools are installed, we can have a look at a simple test example for a possible client of the GET / root endpoint:

```go
func TestConsumerIndex_Local(t *testing.T) {
    // Initialize
    pact := dsl.Pact{
        Consumer: "Consumer",
        Provider: "BookSwap",
    }
    pact.Setup(true)
    // Test case - makes the call to the provider
    var test = func() (err error) {
        baseURL, ok := os.LookupEnv("BOOKSWAP_BASE_URL")
        require.True(t, ok)
        url := fmt.Sprintf("%s:%d/", baseURL, pact.Server.Port)
        req, err := http.NewRequest("GET", url, nil)
        assert.Nil(t, err)
        req.Header.Set("Content-Type", "application/json")
        resp, err := http.DefaultClient.Do(req)
        assert.Nil(t, err)
        assert.NotNil(t, resp)
```

```
    return
  }
  t.Run("get index", func(t *testing.T) {
    pact.AddInteraction().
    Given("BookSwap is up").
    UponReceiving("GET / request").
    WithRequest(dsl.Request{
      Method: "GET", Path: dsl.String("/"),
      Headers: dsl.MapMatcher{
        "Content-Type": dsl.String("application/json"),
      }
    }).
    WillRespondWith(dsl.Response{
      Status: https.StatusOK,
      Body: dsl.Like(handlers.Response{
        Message: "Welcome to the BookSwap Service!",
      }),
    })
    require.Nil(t, pact.Verify(test))
  })
  // Clean up
  require.Nil(t, pact.WritePact())
  pact.Teardown()
}
```

Looking at the client test more closely, we can see that writing a contract test with Pact is not all that different from writing a unit test with Go's standard testing library:

1. The signature of the test is the same as a unit test, conforming to the test name convention and taking in the single `*testing.T` parameter.

2. The Pact DSL is initialized, and we start up the Pact Mock Server using the `Setup()` function. Pact will find a free port on the local machine and then start up the server.

3. We create a test case function that takes in no parameters and returns a single error: `func() error`. This function wraps around the consumer code that calls out to the provider, including setting up any requests required. As we don't have a dedicated client service on the BookSwap application, we simply call out to it using the `http` library.

4. With everything set up, we can run test cases in subtests. This allows us to use the same test techniques that we've seen so far, including the table-driven testing we explored in *Chapter 4, Building Efficient Test Suites*.

5. Inside each subtest, we define a new Pact interaction using the `AddInteraction()` function, which sets up all the prerequisites for contract testing, including starting a Mock Server, if one is running.

6. The `dsl.Interaction` type returned allows us to configure all of the attributes required to describe the contract between the consumer and provider: the request and response body, headers, query parameters, status code, and so on.

7. Once everything has been set up for the test case and expected behavior, we verify that the behavior is as written using the `Verify` function, which takes in the test case that has defined the consumer configuration.

8. Finally, we record the interaction in a file and invoke the `Teardown` function, which stops the Pact Mock Server. By default, Pact will save the contract inside the `pacts` folder in the project.

We can run this test in the same way as we might run any integration test. The output of this test run will be as follows:

```
$ LONG=true go test chapter08/contract_test/consumer_test.go -v
=== RUN   TestConsumerIndex_Local
2023/01/08 16:19:36 [INFO] checking pact-mock-service within
range >= 3.5.0, < 4.0.0
2023/01/08 16:19:36 [INFO] checking pact-provider-verifier
within range >= 1.36.1, < 2.0.0
2023/01/08 16:19:37 [INFO] checking pact-broker within range >=
1.22.3
2023/01/08 16:19:37 [INFO] INFO  WEBrick 1.3.1
2023/01/08 16:19:37 [INFO] INFO  ruby 2.4.10 (2020-03-31)
[x86_64-darwin19]
2023/01/08 16:19:37 [INFO] INFO  WEBrick::HTTPServer#start:
pid=83017 port=52412
=== RUN   TestConsumerIndex_Local/get_index
2023/01/08 16:19:37 [INFO] INFO  going to shutdown ...
2023/01/08 16:19:38 [INFO] INFO  WEBrick::HTTPServer#start
done.
--- PASS: TestConsumerIndex_Local (1.67s)
    --- PASS: TestConsumerIndex_Local/get_index (0.01s)
PASS
ok      command-line-arguments  1.828s
```

The output of the command indicates that the `TestConsumerIndex_Local` test was run against the Pact Mock Server and that it passed. The pact is also written to the `pacts/consumer-bookswap.json` file. This file contains the specified interactions between the consumer and provider, as described by the test.

The consumer has specified the behavior they expect from the provider in the contract specification. Therefore, the provider verification is much simpler:

```go
func TestProviderIndex_Local(t *testing.T) {
  // Initialise
  pact := dsl.Pact{
    Provider: "BookSwap",
  }
  url := getTestEndpoint(t)


  // Verify
  _, err := pact.VerifyProvider(t, types.VerifyRequest{
    ProviderBaseURL: url,
    PactURLs:        []string{PACTS_PATH},
  })
  require.Nil(t, err)
}
```

This simple snippet contains everything required for verification on the provider side:

1. We define the provider verification as a unit test, in the same way as we did on the consumer side.

2. As we run the provider verification against the real service, we do not start the Pact Mock Server, but initialize the Pact DSL.

3. We call the `VerifyRequest` function, passing in the URL to the provider and the path to the consumer-defined contract. This was generated by running the consumer test, as described earlier on.

The URL to the provider and the path to the contract definition have been defined outside the scope of this test, allowing us to run this test in different environments. Once the `BookSwap` application is up and running with the Docker command we saw earlier, we can run the provider verification:

```
$ LONG=true go test chapter08/contract_ test/provider_test.go
-v

=== RUN   TestProviderIndex_Local
2023/01/08 17:46:09 [INFO] checking pact-mock-service within
range >= 3.5.0, < 4.0.0
```

```
2023/01/08 17:46:09 [INFO] checking pact-provider-verifier
within range >= 1.36.1, < 2.0.0
2023/01/08 17:46:09 [INFO] checking pact-broker within range >=
1.22.3
=== RUN    TestProviderIndex_Local/Pact_between__and__
=== RUN    TestProviderIndex_Local/has_status_code_200
   pact.go:638: Verifying a pact between Consumer and BookSwap
Given BookSwap is up GET / request with GET / returns a
response which has status code 200
=== RUN    TestProviderIndex_Local/has_a_matching_body
   pact.go:638: Verifying a pact between Consumer and BookSwap
Given BookSwap is up GET / request with GET / returns a
response which has a matching body
--- PASS: TestProviderIndex_Local (1.52s)
   --- PASS: TestProviderIndex_Local/has_status_code_200
(0.00s)
   --- PASS: TestProviderIndex_Local/has_a_matching_body
(0.00s)
   --- PASS: TestProviderIndex_Local/Pact_between__and__
(0.00s)
PASS
ok      command-line-arguments  1.682s
```

The provider verification passes because the returned response from the BookSwap application is as we have specified on the consumer side. We have now successfully written and run our first contract test with Pact! All of the interaction with the contract testing library has been through a simple Go library, which has also allowed us to write contract tests in the same way as unit tests.

As we have seen, the power of Pact is that it allows developers to easily implement code-first contract tests, so it is definitely a framework that you should consider adding to your projects alongside the practice of contract testing.

> ### The role of the Pact Broker
>
> In the example we have explored, the contract tests were running locally, so they had shared access to the same contract file. However, this is not possible in microservice architectures or consumer-facing services. Teams run a dedicated Pact Broker service that can serve as the URL to the contracts that they wish to write and verify. The Pact Broker can be easily run with Docker using its image available on Docker hub (https://hub.docker.com/r/pactfoundation/pact-broker/).

Breaking up the BookSwap monolith

The discussion in this chapter has been centered around discussing microservice architectures, as distributed systems have become the standard and you will most likely have to work on this kind of system in the near future. However, the BookSwap application is still a monolithic application.

Based on some of the practices we discussed in *Chapter 7, Refactoring in Go*, we can discuss how we might go splitting up the BookSwap monolith. *Figure 8.7* depicts some of the microservices that we could create:

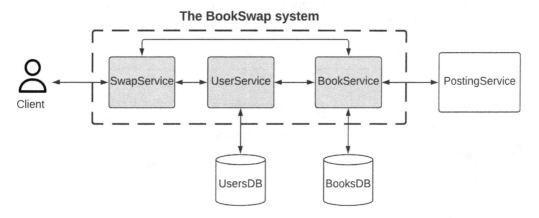

Figure 8.7 – The distributed BookSwap system

The distributed BookSwap system has microservices with well-defined responsibilities:

- SwapService is the entry point to the system and is responsible for handling and routing all the incoming user requests of the system. It has direct dependencies on BookService and UserService, which own the data that SwapService relies on.

- UserService is responsible for all the operations pertaining to user management. The service has persistent storage, UsersDB, which it has full control of inside the system. This storage can take any form, but the service must be able to support the access patterns required by SwapService. This service has a direct dependency on BookService.

- BookService is responsible for all the operations pertaining to book management. This service has its own dedicated persistent storage, BooksDB, which it has full control of inside the system. This service has a direct dependency on PostingService, which is an external service to the BookSwap system.

> **Avoiding a shared database**
>
> `BookService` and `UserService` have been designed to have their own dedicated databases, instead of sharing one single persistent storage. This allows us to enforce data separation between the two microservices, as well as ensure that a database outage does not cause an outage on both of the services.

This simple `BookSwap` system from *Figure 8.7* is the starting point of how we might go about splitting up the `BookSwap` monolith. As we can see, the services have dependencies, so they must support the access patterns required by their consumers. The next step in the monolith splitting process is to design the APIs of the different services.

Figure 8.8 depicts which API calls the services might expose:

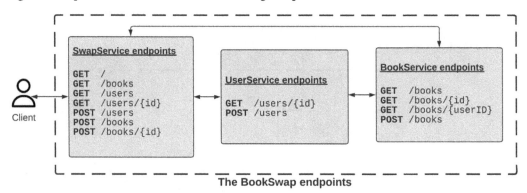

Figure 8.8 – The endpoints of the BookSwap microservices

As previously discussed, `SwapService` is the only user-facing service, with the other services being direct dependencies to it. We can see the following access patterns for the different domains in the `BookSwap` services:

- Books are accessed by their primary ID and by their owner user ID. This access pattern by two indices must be implemented to satisfy the requirements of `SwapService`.

- There is a **one-to-many relationship** between users and books. If we use a SQL database, then the user ID is a foreign key on the books table. This type of dependency can also be implemented in NoSQL tables, even though it feels like a natural fit for SQL databases.

- Users are created and updated using the same `POST` request on the corresponding endpoint. This conforms to RESTful design practices, but this merged operation should be handled lower down on the service level.

- Books are created using a `POST` request but are updated using their ID as the URL parameter. The second update is the implementation of the swap endpoint.

These simple access patterns can be easily implemented with either a SQL or NoSQL persistent storage solution. It is highly recommended that REST endpoints use JSON as content type, especially as JSON marshaling and unmarshaling are natively supported in Go with the `encoding/json` library. We have previously explored persistent storage with PostgreSQL, but most major NoSQL data stores have Go drivers.

The `BookSwap` monolithic application would have lived in one single code base up until now, giving developers full visibility of all the changes that are being made to the application. However, in the microservices world, each service has its own code repository and team ownership.

Figure 8.9 depicts the five service integrations exposed by the new microservice architecture:

Consumer	Provider
Client code	SwapService
SwapService	BookService
SwapService	UserService
UserService	BookService
BookService	PostingService

Figure 8.9 – Five service integrations exposed by the new microservice architecture

The microservices assume the role of consumer and provider according to the flow of data and the request flow between the two services:

- The client code is the consumer that issues the request to `SwapService`, which handles it by relying on the other services of the `BookSwap` application.

- `SwapService` is the consumer as it issues requests to `BookService` and `UserService` in order to process creation and update their corresponding model.

- `UserService` is the consumer and `BookService` is the provider as it fetches the list of books belonging to the user.

- `BookService` is the consumer and the external `PostingService` is the provider, as `PostingService` handles the side effects of all book swaps, which is a critically important detail because these side effects are what deliver the business value of the system in the real world.

Contract testing the integration between `BookService` and the external `PostingService` can help us to validate version upgrades, ensuring that external APIs continue to integrate well with our systems. This is a great way to ensure the continued successful operation of our system and all its dependencies.

As we have seen from the `BookSwap` application in this section, monolithic applications can be converted in to a microservice architecture once the domain and team have the maturity to undertake this journey. In turn, this adds different kinds of complexity to the development, testing, and release processes. That complexity then enables onward scaling of the solution and team. A solid testing strategy, which includes contract testing, can help validate that the microservice architecture is stable, as well as scalable.

Production best practices

The final aspect of the microservice architectures that we will look at is some best practices when it comes to deployment and release. As we previously mentioned, the release pipelines should be automated to make it feasible for teams to release service multiple times a day. In this section, we will briefly explore some common patterns and solutions to consider when migrating to microservice architectures.

Monitoring and observability

In the microservices world, it can be difficult to have an understanding of how data travels through the system and how healthy our system is. This is alleviated by monitoring and observability solutions, which give us the required visibility.

> **Observability versus monitoring**
>
> Observability and monitoring are often used interchangeably, but they have two different intended purposes: observability aims to give teams access to data they need to debug problems, while monitoring aims to track performance and identify service anomalies. This means that monitoring is contained within observability. Observations need to be viewed in terms of meaningful value to the business in order to deliver reliable monitoring of properties, such as availability, performance, and capacity.

We covered some important metrics for performance earlier in this chapter, in the *Performance testing in Go* section. Alongside these important metrics, **structured logging** is often added to track application logs. This type of logging can be analyzed to get an idea of the events that have occurred in our microservice architecture. Some popular open source structured logging libraries are `zap` (`https://pkg.go.dev/go.uber.org/zap`), `logrus` (`https://github.com/sirupsen/logrus`), and `apex/log` (`https://github.com/apex/log`).

Deployment patterns

While a solid test strategy verifies the system for errors and performance issues, no code change or testing strategy is perfect. Deployment patterns will allow us to gradually release changes, making it easier to prevent outages. Two common patterns are as follows:

- **Canary deployments** involve releasing the change to a small percentage of traffic. If the canary is functioning correctly, then we roll out the change to larger percentages of traffic. However, if the metrics recorded in the canary deployment are not positive, we can roll back traffic to the old version of the application, which is still up and running. This minimizes the amount of work that must be done to handle the repercussions of a negative change.

- **Blue-green deployments** involve maintaining two versions of the microservice to be changed. The blue version is running the current version of the service, while the green version is running the updated version. Once the green version has passed testing, user traffic is routed to the green environment. In the case of errors, traffic can be routed back to the blue version. Once the team is confident that the green version is functioning correctly, the blue version can be removed from the running environment or can be used for the next iteration.

These two popular deployment strategies will make it easier to avoid outages when rolling out new versions of a microservice, allowing us to quickly roll back to the previous version in the case of increased error rates. Such strategies are well supported by tools such as Kubernetes and service meshes.

The circuit breaker pattern

The circuit breaker pattern is a development pattern that allows us to avoid **cascading failures**, which is the process of one service increasing the probability that other services will fail. Circuit breakers typically wrap remote calls to other microservices. Once the error rate for calls to the remote services reaches an established threshold, the circuit breaker will immediately fail other requests, allowing the other service space to attempt to recover, and giving users a clear and timely response to explain the situation rather than keeping many requests in flight. An open circuit breaker then retries after a delay, becoming closed and able to pass further requests if the remote services are available, or becoming open again if the problems continue.

A popular open source circuit breaker implementation is the `hystrix-go` (`https://github.com/afex/hystrix-go`) library, which implements error monitoring and retries. This pattern is simple and also requires us to consider default values and fallback behavior for all of our remote calls. The explicit implementation of the error cases for dependency outages brings further resilience to our microservice architecture.

This brings us to the end of our exploration of microservice architecture implementation and testing. As we have seen in this chapter, a comprehensive testing strategy will allow us to take full advantage of the power of microservices, but we must be aware of the difference in the development process in order to be able to efficiently work with microservice architectures.

Summary

In this chapter, we discussed how to test microservice architectures. Having focused on functional testing in previous chapters, we started by exploring non-functional testing. Then, we took a closer look at performance testing, one particularly important type of non-functional testing. Then, we explored the complexities that microservice architectures bring to the development process and learned how contract testing can help with the verification of API integrations.

We learned how to use the Pact tools to write contract tests using the same techniques and processes that developers use for unit testing. Finally, we explored how we might split up the monolithic `BookSwap` application, including which services, endpoints, and contract tests we would write.

In *Chapter 9, Challenges of Testing Concurrent Code*, we will tackle the complex topic of concurrency in Go. We will learn the fundamentals of concurrency in Go and then explore the testing challenges that concurrency introduces.

Questions

1. What is the difference between functional and non-functional testing?

2. What are some key metrics that performance testing should measure?

3. How does performance testing ensure that the system is scalable?

4. What are some of the benefits of microservice architectures? What types of complexity are introduced by microservice architectures?

5. What is contract testing?

Further reading

- *Web Application Security: Exploitation and Countermeasures for Modern Web Applications*, Andrew Hoffman, published by O'Reilly

- *Production-Ready Microservices: Building Standardized Systems Across an Engineering Organization*, Susan J. Fowler, published by O'Reilly

- *Building Microservices: Designing Fine-Grained Systems*, Sam Newman, published by O'Reilly

- *Monolith to Microservices: Evolutionary Patterns to Transform Your Monolith*, Sam Newman, published by O'Reilly

Part 3:
Advanced Testing Techniques

The final part is dedicated to discussing the more challenging aspects of testing complex Go code, as all the tools and techniques required for testing applications are provided by the previous two sections. We begin our exploration by learning about Go concurrency mechanisms and what the concurrency untestable conditions are, including how to use Go's race detector. Then, we revisit and expand our testing of edge cases by making use of fuzz tests and property-based testing, allowing us to test our code with a large amount of input to ensure that it is robust. Finally, we explore how to leverage Go's recently introduced generics capability to write code that can work with different types, learn how to change table-driven tests to verify generic code, and leverage generics to create custom test utilities.

This part has the following chapters:

- *Chapter 9, Challenges of Testing Concurrent Code*
- *Chapter 10, Testing Edge Cases*
- *Chapter 11, Working with Generics*

9

Challenges of Testing Concurrent Code

In the previous chapters, we covered all of the essentials knowledge that TDD practitioners will need to test their applications. We learned how to unit test our code in the development phase, how to integration test our larger components, and how to end-to-end test our entire services. These are essential building blocks for building and running any software project. The test suite allows us to verify that our application is functioning according to the client's requirements.

As the system grows and matures, developers then must inevitably consider how to change and evolve their code, ensuring that their system remains performant and scalable. As discussed in *Chapter 7, Refactoring in Go*, there are some common refactoring techniques that we can use to make the code change process easier. One common system refactoring technique is breaking up monolithic applications and replacing them with microservice architectures. In *Chapter 8, Testing Microservice Architectures*, we learned how to split up the BookSwap application and test the integrations between microservices with the newly introduced technique of contract testing.

As we step into the world of microservice architectures, testing gets more difficult due to two crucial aspects: services are changed by teams within the organization without any central oversight and the operation order can no longer be guaranteed. We covered the integration testing aspects in previous chapters, but we are yet to explore the difficulties brought on by the variations in operation order. Most importantly, we need to explore how to handle the different states that varying operation orders can put an application in.

This chapter will explore the implementation and testing of concurrent code. We will begin by discussing Go's concurrency mechanisms, which are one of the main advantages of the Go programming language. Then, we will explore some common concurrency examples. We will learn how to make use of Go's race detector, which is part of the Go toolchain. Finally, we will discuss what concurrency conditions cannot be tested and see how we can detect concurrency issues in the BookSwap application.

In this chapter, we will cover the following topics:

- Go's concurrency mechanisms – goroutines, channels, and synchronization primitives
- Applied concurrency examples and patterns, including creating thread-safe data structures
- Untestable conditions of concurrent code – race conditions, deadlocks, and starvation
- The usage and limitations of the Go race detector
- Detecting and fixing concurrency issues in the `BookSwap` application

Technical requirements

You will need to have **Go version 1.19** or later installed to run the code samples in this chapter. The installation process is described on the official Go documentation at `https://go.dev/doc/install`.

The code examples included in this book are publicly available at `https://github.com/PacktPublishing/Test-Driven-Development-in-Go/chapter09`.

Concurrency mechanisms in Go

Go's in-built concurrency mechanisms are one of its biggest strengths and are often one of the main reasons developers choose to use Go for their services. Implementing concurrency in Go is easy (and painless!) due to its **goroutines** and **channels**. In this section, we will explore each mechanism and review its behavior so that we can better understand how to use and test them.

Concurrency is a program's ability to process multiple tasks at the same time. This crucial ability allows us to get the most out of the CPU processing power, allowing us to make optimal use of our resources. This is important in all systems in order to be able to process as many requests as possible, without disrupting other flows in the program and keep computing costs low.

Figure 9.1 depicts two concurrent tasks:

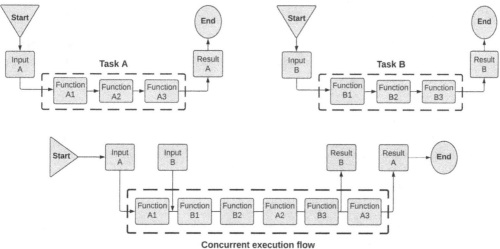

Figure 9.1 – Concurrent execution flow of two tasks

The tasks are divided into functions that form a call stack:

1. In this example, each task is divided into three functions that make up the call stack. The task begins when it receives its input and finishes when it has computed its result or output. **Task A** is divided into three functions: **Function A1**, **Function A2**, and **Function A3**. This separation is the same in **Task B**.

2. The two tasks, **Task A** and **Task B**, are independent of each other. Each task receives its own input and calculates its own result. As the tasks are not connected, they can be computed in any order. This makes them suitable for executing as part of a **concurrent execution flow**.

3. When executing the tasks concurrently, subtasks are **scheduled** and **interrupted** for the most efficient execution. The ability to interrupt the functions in the call stack is a key requirement for the concurrent execution of these two tasks. We will learn how to prevent these interruptions in the following sections.

4. Each task begins when its input is received. In this example, **Input A** is received before **Input B** and its corresponding **Task A** starts execution first.

5. The subtasks, or functions, are executed in an interleaving way, with the CPU executing functions from **Task A** and **Task B** in a combined way. We notice that the subtasks are executed in order within the task. This means that **Function A1** is executed before **Function A2**, but there are **no order guarantees** with regard to timing when it comes to the subtasks of **Task B**.

6. Once the task is completed successfully, the output is returned and the CPU is free to execute other tasks. We notice that even though **Input B** arrives second and **Task B** starts second, it

completes first and **Result B** is returned first. The scheduling of the functions depends on the availability of resources and other factors. We will explore how scheduling works in Go in later sections.

Since there are no order guarantees between concurrently running tasks, we should be careful that the tasks we allow to run concurrently are independent and do not rely on each other. Otherwise, the concurrent execution of the tasks could lead to slow-running tasks or bugs.

> **Avoid ordering assumptions**
>
> Concurrency can be used under the hood in libraries, and it might not always be straightforward to see where it is used. Therefore, we should avoid making assumptions about ordering or execution time. We will learn how to make use of synchronization mechanisms and checks to ensure that conditions are met before execution is started.

Parallelism is often confused with concurrency, but it is a program's ability to execute tasks simultaneously. Unlike concurrency, which does not guarantee task ordering, we know that the task execution in this pattern will be happening in parallel. Tasks should also be independent of each other, as they cannot wait for each other.

Figure 9.2 depicts two parallel tasks:

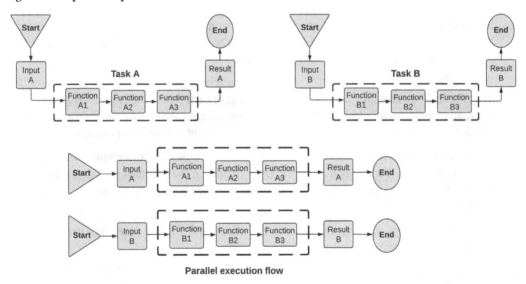

Figure 9.2 – Parallel execution flow of two tasks

The parallel execution flow of two tasks happens simultaneously:

1. The tasks begin executing once **Input A** and **Input B** are received.

2. The tasks are executed simultaneously and independently, without interruption or interleaving.

3. The tasks are completed at the same time, within a margin of error. There will always be deviations in resource usage and performance regardless of how much we attempt to specify them to be identical.

In order to achieve true parallelism, separate computing resources are required. This increases the cost of our system infrastructure, which is undesirable, if not a dealbreaker, for some engineering teams. Therefore, concurrency is often the preferred way to achieve multitasking in programs. As the system becomes successful, properly implemented concurrency can facilitate a smooth transition to parallelism when the system can handle such increased costs.

In Go, the concurrent processing of functions or subtasks is executed using **goroutines**. We will look at what they are, how they are scheduled, and how to synchronize them in the following sections.

Goroutines

Now that we understand the difference between concurrency and parallelism, we will focus our attention on the implementation of concurrency in Go for the remainder of this chapter.

Goroutines are functions or methods that can run concurrently with other functions or methods. They are often referred to as **lightweight threads**, as they have a small memory allocation and run over a much smaller number of OS threads.

It is easy to instruct the Go runtime to run a function in its own goroutine by using the go keyword:

```go
func greet(gr string) {
  fmt.Println("Hello, friend!")
}


func main() {
  go greet()
  fmt.Println("Goodbye, friend!")
}
```

This code snippet creates a main function and a greet function, which takes a string as a parameter and then prints it to the terminal. We instruct the runtime to run the function in its own goroutine by adding the go keyword in front of its invocation. At the end, we print the "Goodbye, friend!" line to signal that the main function has been completed.

We run this little program using the usual command:

```
$ go run chapter09/concurrency/goroutines/main.go
Goodbye, friend!
```

The program does not print the greeting; instead, it only prints the goodbye line. This is due to the behavior of programs and goroutines. *Figure 9.3* presents a visualization of these properties:

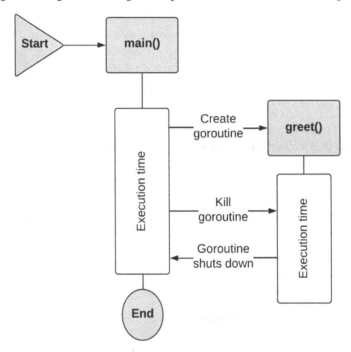

Figure 9.3 – Goroutine execution of the greeting program

The program does not print the greeting to the terminal because of the intended **non-blocking behavior** of goroutine creation:

1. The main function starts when we run our program. This function runs in its own goroutine, which we will refer to as the **main goroutine**. The main function has its own **execution time** based on the statements contained inside the body of the main function.

2. During main function execution, the main goroutine instructs the Go runtime to **create a goroutine** and run the greet function in this goroutine. This main goroutine has a **parent-child relationship** with the main goroutine. We will refer to this child goroutine as the **greet goroutine**.

3.   The creation of the child goroutine, which will run our greet function, is a **non-blocking operation**. This allows us to achieve the multitasking aspect of concurrency that we previously discussed.

4.   As the main goroutine is not blocked, it finishes its own work and completes its execution time. Once the main goroutine has completed, the Go runtime cleans up all of its resources. As the main goroutine has a parent-child relationship with the greet goroutine, the runtime **kills the greet goroutine**.

5.   The greet goroutine immediately stops execution and **shuts down**. Depending on how much execution time it has received from the CPU, the greet goroutine may be able to execute its print to the terminal or not.

Due to these properties, the program does not manage to reliably print the greeting to the terminal. We need to stop the main goroutine from shutting down in order to give time for the child goroutine to finish its execution.

One solution is to block the main goroutine from terminating by invoking the time.Sleep function for a predetermined amount of time, such as 1 second. Another, more interesting, solution is to signal that the greet goroutine has completed its work by writing a value to a shared variable:

```go
var finished bool
func greet() {
  fmt.Println("Hello, friend!")
  finished = true
}


func main() {
  go greet()
  for !finished {
    fmt.Println("Child goroutine not finished.")
    time.Sleep(10 * time.Millisecond)
  }
  fmt.Println("Child goroutine finished.")
  fmt.Println("Goodbye, friend!")
}
```

The two functions share memory space, so it is possible for them to write and read to shared variables. The code snippet demonstrates this:

1.   We create a variable of the bool type at the top, named finished. The purpose of this variable is to provide a signal to the main function that the greet function has completed.

2. Once the greet function writes its greeting to the terminal, it sets the value of the `finished` variable to `true`.

3. Inside the body of the main function, we create a `for` loop, which will execute until the value of the `finished` variable is `true`. Using the `time.Sleep` function, we poll the value of the variable every 10 milliseconds.

4. Once the `for` loop completes, the main goroutine completes its execution and all of the resources of both goroutines are cleaned up.

Running this program will print the following:

```
$ go run chapter09/concurrency/goroutines/main.go
Child goroutine not finished.
Hello, friend!
Child goroutine finished.
Goodbye, friend!
```

Finally, using this simple approach of writing to a shared variable, we have managed to block the main goroutine until its child goroutine finishes. We are finally able to see the greeting printed in the terminal and the program is executing correctly.

This way of sharing information between goroutines is known as **communicating by sharing memory** and it is the traditional way to deal with concurrency in other programming languages. This approach is not without drawbacks. However, Go has another approach, which we will explore in the next section.

Channels

Channels provide another way for goroutines to communicate with one another. We can think of the in-built channel type as a pipe through which we can safely send information between goroutines, without having to use shared variables or memory. In Go, this principle is known as **sharing memory by communicating**.

Figure 9.4 depicts the main operations and syntax of channels:

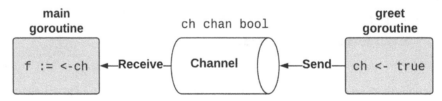

Figure 9.4 – Operations and syntax of Go channels

The interaction with the channel demonstrates the syntax of its two main operations:

1. `ch chan bool`: The channel is a built-in type in Go, so it does not require us to import any libraries. A channel is declared using the `chan` keyword followed by a data type, `bool`, which the channel will be able to transport. Only this type of variable can be transported through it. This is compiler enforced.

2. `ch <- true`: The first operation that channels support is the **send operation**. The **channel operator** looks like an arrow, `<-`, and indicates the way that the data is flowing through the channel. In this case, the arrow points toward the channel where we send a `true` value to it.

3. `f := <-ch`: The counterpart to the send operation is the receive operation. This operation is performed by pointing the channel operator away from the channel and allocating the receive to a local variable named `f`.

This is the basic usage of channels, although we will explore some further subtleties in later sections. The send and receive operations are **blocking and synchronous**, so both parties of the transaction need to be available for the operations to complete.

Channels are a great **synchronization and communication mechanism**. We can make use of them to synchronize our main and greeter goroutines using more concise code:

```go
func greet(ch chan bool) {
  fmt.Println("Hello, friend!")
  ch <- true
}
func main() {
  ch := make(chan bool)
  go greet(ch)
  <-ch
  fmt.Println("Child goroutine finished.")
  fmt.Println("Goodbye, friend!")
}
```

This simplified version of the solution makes use of a channel to synchronize the two goroutines:

1. The `greet` function is changed to take in a channel parameter. Similarly, for maps and slices, the channel type has a built-in pointer reference, so we do not need to explicitly pass it by pointer using the `&` operator.

2. Once the greeting is printed, the `greet` function sends the `true` value to the channel. This will signal to the main goroutine that the `greet` function has successfully completed.

3.   Inside the main function, we initialize a channel using the `make` function. The zero value of the channel type is `nil`, so we use the `make` function to create a channel that is ready to use. Under the hood, the `make` function will allocate all the resources required.

4.   Once the `greet` function has been started in its own goroutine, the main function invokes the receive operation on the channel. Since the send and receive operations on channels are blocking, this will block the main goroutine until the `greet` goroutine has completed and has been able to send the value through the channel.

The use of channels has simplified the implementation by removing the need to poll for the value of the `finished` variable. We also notice that the channel variable, `ch`, has been initialized inside the main function and been passed as a parameter. Since there is now no global variable, we have removed the need to communicate by sharing memory between the two goroutines.

Channels support one final operation, the close operation. Unlike sends and receives, closing a channel changes the state of the channel and indicates the completion of work to the channel's receivers. It is an operation that can be used for the purpose of synchronization, as opposed to supporting information exchange and communication between goroutines. A closed channel will immediately return the zero value of the channel type to all receive operations and cause a panic on all send operations that are attempted on the channel in the future.

Since the purpose of our channel is to synchronize the `greet` and `main` goroutines, we can make use of the close operation to further simplify our code:

```go
func greet(ch chan struct{}) {
  fmt.Println("Hello, friend!")
  close(ch)
}

func main() {
  ch := make(chan struct{})
  go greet(ch)
  <-ch
  fmt.Println("Child goroutine finished.")
  fmt.Println("Goodbye, friend!")
}
```

There are a few tweaks we have made to our solution. The data type of the channel is now the empty `struct{}`, which reduces the memory footprint of the channel. Inside the `greet` function, we close the channel immediately after the function prints its greeting. While these changes do not seem significant, this solution will work for signaling to multiple receivers that work has completed, as opposed to having to write multiple values on the channel. This is a powerful mechanism that we can make use of to solve a variety of problems.

Figure 9.5 summarizes the behavior of the channels we have studied so far:

	Nil channels	Initialised channels	Closed channels
Send	Block	Block until a receiver arrives	Panic
Receive	Block	Block until a sender arrives	Complete immediately with zero value of data type
Close	Panic	Complete immediately	Panic

Figure 9.5 – Summary of channel operations and states

This figure is a useful reference for understanding how channels will behave in our code:

1. **Nil channels** are channels that have not been correctly initialized using the `make` function. They cannot be used to send information but are useful for passing to goroutines when those goroutines are started. The nil channel will be initialized for use at a future time:

    I.   Send operations will block until the channel is initialized, after which the rules for initialized channels apply.

    II. Receive operations behave identically to send operations.

    III. Close operations panic on nil channels. As nil channels are not ready to send information through, it would not make sense to close them. It is therefore considered a fatal error if we attempt to close nil channels.

2. **Initialized channels** are created using the make function and are ready to be used. They are ready for sending information through:

    I.   Send operations will block until a receiver arrives. The sending goroutine will not be able to execute past this point until the operation completes.

    II. Receive operations will block until a value arrives from the sender. As sends and receives are synchronous operations, both goroutines must be ready to complete the operation for the two parts of the transaction to be completed. So, if the sender starts up but the receiver is not yet ready, this will mean the sender halts until the receiver is ready, which can be a helpful property.

III.   Close operations complete immediately. Once the first operation completes, the channel will move into the **Closed Channel** state.

3.   Closed channels are initialized channels that have been successfully closed. Channels in this state signal that they will no longer be able to transport information:

I.   Send operations will panic. There is no easy way to know whether a channel is closed, so the panic lets senders know that they should stop sending values to it, but you should code carefully in order to avoid encountering a panic.

II.   Receive operations will immediately complete with the zero value of the channel's data type. As we have seen in our greeter example, we can use the receive operation on closed channels as a synchronization mechanism.

III.   Close operations will panic, as channels can only move into the closed state once. Again, defensive coding (for example, the single responsibility principle where only one part of your code is responsible for closing the channel) can help to control this.

One final aspect to note is that once a channel is closed, it cannot be opened again. This can create some complications when using them when solving more complicated problems. Now that we understand the fundamental behavior of goroutines and channels, we can explore some commonly applied concurrency examples in the next section.

Applied concurrency examples

So far, we have learned about the main operations and behavior of goroutines and channels. These two concurrency mechanisms are important to understand, as they are pivotal to how Go implements concurrency. However, the Go standard library also includes concurrency primitives in its `sync` package. It contains synchronization primitives with a broad variety of uses:

- `sync.Map` is a map implementation that is safe for concurrent use. We will explore how to create other thread-safe data structures in the next section.

- `sync.Mutex` is an exclusion lock. It allows us to gatekeep resources for usage by one goroutine at a time. It is also possible to take a read-only or a read-write mutex depending on the problem being solved.

- `sync.Once` is a specialized lock that can only be acquired once. This is useful for wrapping around statements, such as cleanup code, which should only run once.

- `sync.Pool` is a temporary set of objects that are individually saved and retrieved. It can be seen as a cache of objects, making it easy to create thread-safe lists.

- `sync.WaitGroup` waits for a collection of goroutines to finish. This primitive has a counter and a lock under the hood, allowing it to keep track of how many goroutines it will need to wait for before completing. This can greatly simplify a main goroutine.

You can read the full documentation of the synchronization primitives of the `sync` package in the official Golang documentation (`https://pkg.go.dev/sync`). These well-designed synchronization primitives give us the tools to solve many types of problems. Let us have a look at some of them in action in the next sections.

Closing once

As we saw in *Figure 9.5*, channels panic if we attempt to close a channel multiple times. This is a great candidate for using `sync.Once`, although we can imagine other great uses of this mechanism, such as implementing the **Singleton pattern** or executing clean-up functions.

This specialized lock is easy to use to ensure that a channel is only closed once:

```go
func safelyClose(once *sync.Once, ch chan struct{}) {
    fmt.Println("Hello, friend!")
    once.Do(func() {
        fmt.Println("Channel closed.")
        close(ch)
    })
}
func main() {
    var once sync.Once
    ch := make(chan struct{})
    for i := 0; i<3; i++ {
        go safelyClose(&once, ch)
    }
    <-ch
    fmt.Println("Goodbye, friend!")
}
```

We implement the safe closing of a channel by wrapping it around the close operation:

1. We create the `safelyClose` function, which takes in a pointer to the `sync.Once` type and the channel created by the main function. Note that unlike the channel type, we need to pass the `Once` type using explicit parameter pointer types.

2. Inside the `safelyClose` function, we call the close operation on the channel inside the `once.Do` method. The Do method takes a function as a parameter, so we wrap our statement inside an anonymous function.

3. Inside the main function, we create a zero-value `sync.Once` instance. There is no special initialization we need to undertake with synchronization primitives, so the zero value is ready for use.

4. We create multiple goroutines that execute the `safelyClose` function using a `for` loop. These goroutines all share the same `once` and channel instances.

5. Finally, we block the main goroutine with a receive operation from the channel. This operation will complete as soon as the first goroutine closes the channel.

Running the example program shows that multiple goroutines are started, but the channel is closed only once:

```
$ go run chapter09/concurrency/once/main.go
Hello, friend!
Channel closed.
Hello, friend!
Goodbye, friend!
Hello, friend!
```

As we can see from the output, multiple goroutines are started, but the channel is only closed once. `sync.Once` is simple to use, but it can help us build safety around operations that should only be executed once, such as closing a channel.

Thread-safe data structures

Another frequent problem that engineers solve is building **thread-safe data structures**. These types of structures are safe for reading and writing by multiple goroutines. By default, Go's slices and maps are not safe for concurrent use, so we will need to be mindful of multiple goroutines accessing shared data structures and resources. This is one of the reasons why communicating using channels, which are thread safe, is preferred to communicating via shared memory, represented by data structures or variables.

`sync.Map` (`https://pkg.go.dev/sync#Map`) is an implementation of a map that is thread safe. This map uses locks under the hood, so it will not be as performant as the built-in map type. The synchronized map exposes methods providing reading and writing functionality:

```
const workerCount = 3
func greet(id int, smap *sync.Map, done chan struct{}) {
    g := fmt.Sprintf("Hello, friend! I'm Goroutine %d.", id)
    smap.Store(id, g)
    done <- struct{}{}
}
```

```
func main() {
  var smap sync.Map
  done := make(chan struct{})
  for i := 0; i < workerCount; i++ {
    go greet(i, &smap, done)
  }
  for i := 0; i < workerCount; i++ {
    <-done
  }
  smap.Range(func(key, value any) bool {
    fmt.Println(value)
    return true
  })
  fmt.Println("Goodbye, friend!")
}
```

We interact with the synchronized map through wrapper methods:

1. We declare the `workerCount` constant at the top of the program, which will denote the number of goroutines we will be starting.

2. The `greet` function takes in three parameters: an ID, a pointer to `sync.Map`, and a channel for us to signal that the goroutine has finished its work. We format a greeting string, which makes use of the ID that was passed in. Then, we save it in the map using the `Store` method and write a value to the `done` channel to signal to the main goroutine that this worker goroutine has finished.

3. Inside the main function, we initialize the map. The zero value of this map is ready for usage, just as we saw with `sync.Once` earlier. We also initialize a channel, which we will use to signal to the main goroutine that the worker goroutines have completed.

4. Then, we run through two `for` loops. The first loop starts the `greet` function in its own goroutine, while the second waits until values are received on the `done` channel. This allows us to wait for all goroutines to complete before continuing.

5. Finally, we read all the values contained in the map using the `Range` method, which takes an anonymous function as a parameter. We print the entries and return `true`, which will allow the `Range` method to continue looping.

The output of this program shows that the greetings can be saved and retrieved concurrently:

```
$ go run chapter09/concurrency/syncmap/main.go
Hello, friend! I'm Goroutine 2.
Hello, friend! I'm Goroutine 0.
Hello, friend! I'm Goroutine 1.
Goodbye, friend!
```

The built-in map type will panic when written to by multiple goroutines, so you should make sure to use the synchronized map in this case.

Similar to the approach of sync.Map, we can create our own thread-safe custom data structures by using the sync.Mutex lock to limit access to the underlying data structure. For example, we can create a thread-safe **Last In First Out (LIFO)** stack by following this approach:

```go
// Thread safe LIFO Stack implementation
type Stack struct {
  lock sync.Mutex
  data []string
}
// Push adds the given element to the end of the list
func (s *Stack) Push(el string) {
  defer s.lock.Unlock()
  s.lock.Lock()
  s.data = append(s.data, el)
}
// Pop removes and returns the last element from the list,
// or an error if the list is empty.
func (s *Stack) Pop() (*string, error) {
  defer s.lock.Unlock()
  s.lock.Lock()
  if len(s.data) == 0 {
    return nil, fmt.Errorf("stack is empty")
  }
  last := s.data[len(s.data)-1]
  s.data = s.data[0 : len(s.data)-1]
  return &last, nil
}
```

The stack implementation makes use of `sync.Mutex`, which exposes two methods, `Lock` and `Unlock`, to limit access to the underlying data slice:

1. The custom `Stack` struct has two fields, a lock and a data slice. These are unexported fields, as they should only be managed by the stack data structure itself.

2. `Stack` has two methods. `Push` adds the element to the end of the data slice, while `Pop` removes the last element from the data slice and returns it. If the slice is empty, then the `Pop` method will return an error.

3. Both functions make use of the lock that is of the `sync.Mutex` type to ensure that both methods are called by one goroutine at a time. We make use of the `defer` keyword to ensure that the lock is released regardless of which execution path the method goes through.

`sync.Mutex` is a versatile locking mechanism that can be used to block access to any code segment that accesses shared resources or requires unique control of a resource. This is known as a **critical code section**.

Similarly, the `sync` package also provides `sync.RWMutex`, which provides control of locking reads and writes separately. This level of control may be useful for creating thread-safe data structures that are used by many goroutines.

Waiting for completion

The final synchronization primitive that we will explore in this section is `sync.WaitGroup`. Under the hood, `WaitGroup` manages an inner counter that maintains how many resources are left to complete. This specialized lock allows us to wait for multiple goroutines to complete, allowing us to simplify our synchronized map example from the previous section:

```
const workerCount = 3
func greet(id int, smap *sync.Map, wg *sync.WaitGroup) {
  defer wg.Done()
  g := fmt.Sprintf("Hello, friend! I'm Goroutine %d.", id)
  smap.Store(id, g)
}
func main() {
  var smap sync.Map
  var wg sync.WaitGroup
  wg.Add(workerCount)
  for i := 0; i < workerCount; i++ {
    go greet(i, &smap, &wg)
  }
```

```
  wg.Wait()
  smap.Range(func(key, value any) bool {
    fmt.Println(value)
    return true
  })
  fmt.Println("Goodbye, friend!")
}
```

We have made a few key changes that greatly simplify our solution:

1. The `greet` function takes in a pointer to `sync.WaitGroup` instead of the done channel. At the top of the function, we defer the `Done` method on `WaitGroup`, which decreases its inner counter by 1, signaling that this goroutine has completed.

2. Inside the main function, we initialize `sync.WaitGroup`, which is ready for use. We add `workerCount` to the inner counter, signaling to it how many goroutines we will start. `WaitGroup` will block until this inner counter reaches zero, which will happen as each child goroutine calls the `Done` method once as it finishes.

3. Finally, we invoke the `Wait` method further down in the `main` function. This method will block until the inner counter of `WaitGroup` reaches 0. This removes the need to read messages from our channel for each completed goroutine inside a `for` loop.

This brings us to the end of our exploration of Go concurrency fundamentals and applications. As we have seen, Go concurrency makes use of goroutines, channels, and synchronization primitives. We can easily use locks to create thread-safe data structures and ensure that critical code sections are only accessed one goroutine at a time. In the next section, we will learn what problems the newly introduced aspect of concurrency brings to our programs.

Issues with concurrency

Writing concurrent code in Go is elegant and simple. However, it does make our code more complex. Developers need to be familiar with the behavior of concurrency mechanisms to understand the code they are reading. Furthermore, as timing plays a crucial part in how goroutines behave, we might have a hard time reproducing potential bugs. In this section, we look at three common concurrency issues. As we deep dive into each example, we will also have the opportunity to understand the behavior of Go's concurrency mechanisms.

Data races

A **data race** is the most common concurrency issue. This issue occurs when multiple goroutines access and modify the same shared resource concurrently. This is one of the reasons why we should avoid sharing the state between goroutines, preferring to share information between goroutines using channels.

We modify our previous greeting example by saving the formatted greetings into a slice, instead of immediately printing the greeting to the terminal:

```
const workerCount = 3
var greetings []string
func greet(id int, wg *sync.WaitGroup) {
  defer wg.Done()
  g := fmt.Sprintf("Hello, friend! I'm Goroutine %d.", id)
  greetings = append(greetings, g)
}
func main() {
  var wg sync.WaitGroup
  wg.Add(workerCount)
  for i := 0; i < workerCount; i++ {
    go greet(i, &wg)
  }
  wg.Wait()
  for _, g := range greetings {
    fmt.Println(g)
  }
  fmt.Println("Goodbye, friend!")
}
```

At first glance, the code example hasn't been modified very much:

1. At the top of the program, we declare the greetings string slice that we will be saving the greetings into. We also declare the workerCount constant as 3, which is how many goroutines we will be running.

2. The greet function takes in two parameters, the goroutine ID and a pointer to sync. WaitGroup. At the end of the function, we append the formatted greeting, g, to the greetings slice.

3. In the main function, we create sync.WaitGroup and run the greet function in multiple goroutines. WaitGroup is used to ensure that the main goroutine waits for all of its worker goroutines to complete. At the end of the main function, once all the greet goroutines have completed, we loop over the greetings slice and print each entry to the terminal.

As the `main` function waits for all the goroutines to complete, we expect that all goroutines will have their greeting saved correctly. As `workerCount` is equal to 3, we expect three lines to be printed to the terminal. Let us run this program in the usual way and see its output:

```
$ go run chapter09/concurrency/data-races/main.go
Hello, friend! I'm Goroutine 2.
Hello, friend! I'm Goroutine 1.
Goodbye, friend!
```

Looking at the output, we see only two goroutines have had their results recorded. We can see that something has gone wrong with the code changes we have made.

This code example suffers from a data race. *Figure 9.6* depicts the sequence of events happening in this example:

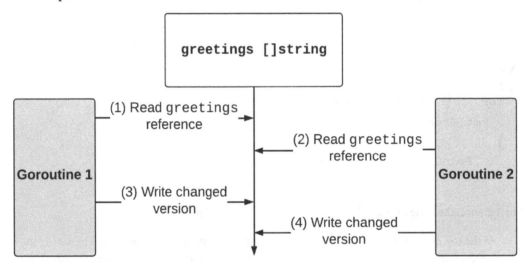

Figure 9.6 – The data race events

As multiple goroutines attempt to append their results to the `greetings` slice, they actually perform a few operations under the hood:

1. **(1) Read `greetings` reference**: **Goroutine 1** begins its execution by reading the reference to the greetings slice. It will complete its operations based on this value.

2. **(2) Read `greetings` reference**: At a later time, **Goroutine 2** begins its execution by reading the reference to the greetings slice. This may or may not be the same value that **Goroutine 1** has read.

3. **(3) Write changed version**: During its execution, **Goroutine 1** is ready to write its changes and complete its execution. If there is space in the underlying array, the element is appended

to it. Otherwise, a new, larger array is created and the elements are copied to it. A new slice is created with a reference to the updated underlying array.

4. **(4) Write changed version**: Finally, **Goroutine 2** is ready to write its changes as well. However, it is unaware of any changes that **Goroutine 1** has made up to this point. It is still working based on the reference that it has read at point **2**. **Goroutine 2** writes its changes, overwriting all the work that **Goroutine 1** will have saved at point **3**.

As the `greetings` slice is not protected by locks, goroutines can be interrupted at any point in this process. As these changes interleave, the goroutines can end up overwriting each other's changes, leading to an inconsistent result. Depending on timing, your output may well look different from the preceding result. Also depending on the timing, we might see all of the greetings printed to the terminal and assume the program is functioning correctly, or we might see the inconsistent behavior we saw during our test run. Data races are commonly occurring issues in the world of concurrency and they can be hard to find and replicate.

Deadlocks

Deadlocks are another common concurrency issue. This issue occurs when goroutines are blocked waiting for a resource that never becomes available. The goroutines will never be able to proceed. The Go runtime will detect when your program becomes blocked and trigger a panic, shutting down and cleaning up resources.

To fix our data race from the previous section, we'll modify the code to make use of a channel to allow only one goroutine at a time to append to the greetings slice:

```go
var greetings []string
const workerCount = 3
func greet(id int, ch chan struct{}, wg *sync.WaitGroup) {
  defer wg.Done()
  g := fmt.Sprintf("Hello, friend! I'm Goroutine %d.", id)
  <-ch
  greetings = append(greetings, g)
  ch <- struct{}{}
}
func main() {
  ch := make(chan struct{})
  var wg sync.WaitGroup
  wg.Add(workerCount)
  for i := 0; i < workerCount; i++ {
    go greet(i, ch, &wg)
```

```
  }
  ch <- struct{}{}
  wg.Wait()
  for _, g := range greetings {
    fmt.Println(g)
  }
  fmt.Println("Goodbye, friend!")
}
```

At first glance, the example seems reasonable:

1. The greet function takes in three parameters: an ID, a channel, and a pointer to WaitGroup. Inside the function, we read from the channel, append our greeting to the greetings slice, then write to the channel.

2. Inside the main function, we initialize the channel and WaitGroup. These are the synchronization mechanisms that our goroutines will use.

3. We then write a for loop, which will start as many goroutines running the greet function as workerCount, which is three.

4. After the loop, we send a value to the channel to get the first goroutine started. It also signals to the worker goroutines that the main goroutine is ready to process their results.

This seems like a reasonable technical solution that could ensure our data races are fixed. Let us run this program in the usual way and see its output:

```
$ go run chapter09/concurrency/deadlock/main.go
fatal error: all goroutines are asleep - deadlock!
goroutine 1 [semacquire]:
sync.(*WaitGroup).Wait(0x0?)
        /usr/local/go/src/sync/waitgroup.go:139 +0x52
main.main()
        .../chapter09/deadlock/main.go:28 +0xd5
goroutine 19 [chan send]:
main.greet(0x0?, 0x0?, 0x0?)
        .../chapter09/deadlock/main.go:17 +0x165
created by main.main
        .../chapter09/deadlock/main.go:25 +0x4f
exit status 2
```

This program suffers from a deadlock, which is detected by the Go runtime. The stack trace indicates that two goroutines are blocked:

- The main goroutine cannot complete the `Wait` method of `WaitGroup`

- One of the worker goroutines cannot complete its channel send operation

This deadlock is caused by the synchronous nature of channel operations. The last worker goroutine tries to send to the channel and signal that its work has completed, but there is no remaining receiver on the channel. Due to this, the worker remains blocked, `WaitGroup` never unblocks, and the whole program freezes.

Common causes for goroutines becoming blocked are waiting to complete channel operations or waiting for one of the locks in the `sync` package to become available. Understanding the behavior of the concurrency mechanisms we are using is the best tool for avoiding issues and bugs.

Buffered channels

By default, channels are **unbuffered**, meaning that they have no capacity to store or buffer values. This is why all the channel operations that we have seen so far have been **synchronous**. However, this can be limiting for senders and receivers that operate at different speeds. A special kind of channel addresses this limitation.

Buffered channels have the capacity to accept a predefined, limited number of values without a receiver. This allows us to process a limited number of **asynchronous** operations. The capacity of a channel is predefined, at initialization, with an optional parameter to the `make` function:

```
ch := make(chan Type, capacity)
```

The capacity is an integer, which has a default value of 0 for unbuffered channels. This parameter defines the size of the backing array that will save the channel's values.

Figure 9.7 depicts the send and receive operations on the two types of channels:

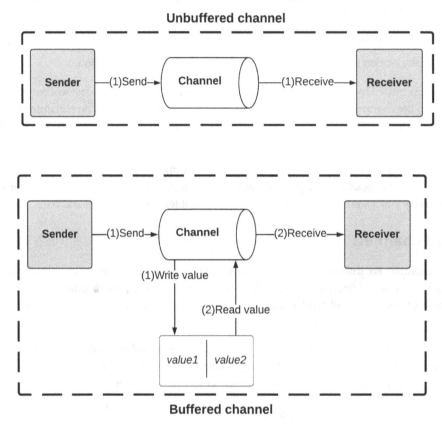

Figure 9.7 – Send and receive operations on channels

The timing of operations is the key difference between the channels:

- On **unbuffered channels**, both the send and receive operations happen at the same time. The channel does not store any values and can only complete the operation once both the sender and receiver are available.

- On **buffered channels**, the channel has a limited capacity buffer that can save values, if it has the capacity to do so. The send and receive operations complete at different times, as the channel saves the sender's value in its buffer. Once the receiver is ready, it can read the available value from its buffer and pass it on to the receiver.

- When the buffer is at capacity, buffered channels will block send operations, behaving like an unbuffered channel until the buffer starts to be emptied by the receiver.

We can make use of buffered channels to allow the `greet` workers to complete as soon as they write their value, instead of waiting for the main goroutine to be available to receive their values:

```go
const workerCount = 3
func greet(id int, ch chan string) {
  g := fmt.Sprintf("Hello, friend! I'm Goroutine %d.", id)
  ch <- g
  fmt.Printf("Goroutine %d completed.\n", id)
}
func main() {
  ch := make(chan string, workerCount)
  for i := 0; i < workerCount; i++ {
    go greet(i, ch)
  }
  fmt.Println(<-ch)
  fmt.Println(<-ch)
  fmt.Println("Goodbye, friend!")
}
```

This simple example demonstrates the usage of buffered channels:

1. The `greet` function takes in two parameters again: an ID and channel with the `string` data type. The buffered channel has the same type as the unbuffered channel, so the `greet` function cannot detect whether it is using a buffered or unbuffered channel.

2. Inside the `greet` function, we format the greeting and send it to the channel.

3. At the top of the `main` function, we initialize the buffered channel by passing `workerCount` as the capacity of the channel. Then, we start the greet function in its own goroutine inside the `for` loop, passing the index and the channel as the parameters of the function.

4. Finally, we print and receive two values from the channel and terminate the program.

We run the program in the usual way to see how it behaves:

```
$ go run chapter09/concurrency/buffered-channels/main.go
Goroutine 1 completed.
Goroutine 2 completed.
Goroutine 0 completed.
Hello, friend! I'm Goroutine 2.
Hello, friend! I'm Goroutine 1.
Goodbye, friend!
```

The program functions as intended: the worker goroutines complete immediately and the main goroutine prints two messages to the terminal, then completes successfully. However, this program does have an issue. The third greeting of the greeter goroutine is successfully sent to the channel but is never received. From the point of view of the greeter, its result was correctly sent and processed, when in fact the main goroutine never processed it.

Since the receiver is only ready twice, our program has a **leaked resource**, which is a resource that has not been released correctly. While the Go garbage collector will collect unused memory, we should avoid writing this kind of code, as it can cause issues and bugs if these operations are performed at scale.

Buffered channels have a limited capacity to ensure that these types of resource leaks are avoided. They are often used to implement the **worker pool concurrency pattern**, which is the implementation of a collection of goroutines waiting to repeatedly process jobs.

So far, we have discussed the behavior and issues of concurrency mechanisms by studying code examples and reasoning around the problems that we are able to reproduce. In the next section, we will discuss how to make use of the Go tools to detect concurrency issues in our programs.

The Go race detector

In *Chapter 8, Testing Microservice Architectures*, we explored how to use the pprof tool to profile the CPU and memory usage of Go applications. One of the essential tools that can help us find issues with concurrency is the Go race detector. It is a powerful tool that analyzes our code to find concurrency problems when an application is running.

Go's race detector was added to the toolchain in Go 1.1, released in 2012. This tool was designed to help developers find race conditions in their code. As we have seen in the previous examples, writing concurrent code in Go is easy, but bugs can appear in even the most readable and well-designed code.

The race detector is enabled using the -race command-line flag, alongside the go command. For example, we can instruct it to run alongside our program:

```
$ go run -race main.go
```

The race detector can be used with other commands as well, including the build and test commands. This makes it easy to use the detector to find data races in your application at any stage in the development process.

Once the detector is enabled, the compiler records memory access and the Go runtime analyzes these records for data races. As we know, data races are typically caused by multiple goroutines accessing and modifying one shared resource without making use of synchronization mechanisms.

When a data race occurs, the detector will print a report with details of the problem, pinpointing the problem and guiding an observant developer toward the fix for the detected issue. Let us try it out with our data race example from the previous section:

```
$ go run -race chapter09/concurrency/data-race/main.go
==================
WARNING: DATA RACE
Read at 0x0000011e6d70 by goroutine 8:
  main.greet()
      .../chapter09/data-races/main.go:15 +0xf5
Previous write at 0x0000011e6d70 by goroutine 7:
  main.greet()
      ../chapter09/data-races/main.go:15 +0x1b3
==================
==================
WARNING: DATA RACE
Read at 0x00c00009e000 by goroutine 9:
  runtime.growslice()
      /usr/local/go/src/runtime/slice.go:178 +0x0
  main.greet()
      .../chapter09/data-races/main.go:15 +0x12f
Previous write at 0x00c00009e000 by goroutine 7:
  main.greet()
      .../chapter09/data-races/main.go:15 +0x164
==================
Hello, friend! I'm Goroutine 0.
Hello, friend! I'm Goroutine 2.
Goodbye, friend!
Found 2 data race(s)
exit status 66
```

As expected, the race detector finds some issues with our data race example. The output points out the problem:

1. The first data race is detected in the greet function at main.go:15. One goroutine reads a variable, while another goroutine writes to it.

2. The second data race is happening as the slice grows during append, which is indicated by the call to runtime.growslice(). This function copies the slice and handles the allocation

of a larger backing array, if it is required. The modifications to this slice are also happening in an interleaving manner, with reads and writes happening in different goroutines.

3. Finally, the output of the program is printed and the race detector summarizes that two data races have been found.

As we already suspected, the concurrent changes made to the shared slice without synchronization mechanisms have caused a data race. The code block identified by the race detector is as follows:

```
func greet(id int, wg *sync.WaitGroup) {
  defer wg.Done()
  g := fmt.Sprintf("Hello, friend! I'm Goroutine %d.", id)
  greetings = append(greetings, g)
}
```

The line highlighted by the detector is the read and write to the greetings slice during the append function. As we discussed in previous sections, the append function consists of multiple operations, which can cause data races if they are interleaving across multiple goroutines.

Due to the instrumentation that the race detector requires, it is only able to find data races as they are triggered. Therefore, our application should be subjected to realistic workloads and user journeys in order to detect issues.

According to the official Go documentation (`https://go.dev/blog/race-detector`), race-enabled applications use 10 times the CPU and memory, so we should avoid running them in production. Instead, we should run our load tests and integration tests with the race detector enabled, since these tests usually exercise the most important parts of the program.

> **Limitations of the race detector**
> While the race detector is a very useful tool, we should remember that it can only check for race conditions. While it does not flag any false positives, the code can contain other concurrency issues. We should bear in mind that the race detector is simply an indicator.

Untestable conditions

While the Go race detector is a useful tool, concurrency testing is difficult to perform and prove to be correct. The race detector only focuses on finding data races, but we have already seen that there are other concurrency issues in the previous section, *Issues with concurrency*, where we explored deadlocks and leaked resources.

Due to the dependency on timing, there are four essentially untestable concurrency problems:

1. **Race conditions**: Unstable or inconsistent behavior due to multiple goroutines that read and modify a shared resource without the correct usage of synchronization mechanisms. For example, goroutines reading and incrementing a common counter.

2. **Deadlocks**: Goroutines becoming blocked waiting for resources that never become available, either because they never reach the required state or because another goroutine has locked the resources and never released them. For example, a goroutine is waiting to receive from a nil channel, which never becomes initialized.

3. **Livelocks**: Similar to deadlocks, goroutines become livelocked when they continue to attempt to acquire resources that never become available, either because they never reach the required state or because another resource has locked the resources and never released them. In this case, goroutines will waste CPU continuing to retry impossible operations. For example, a goroutine periodically polls to write to a variable that has been locked by another goroutine, which is waiting for a resource that the first goroutine has locked and never received.

4. **Starvation**: Similar to livelocks, goroutines cannot get all the resources needed to continue processing. One or more goroutines are prevented from doing meaningful work by greedy goroutines that do not release resources. For example, a goroutine locks a resource and then proceeds to execute a very long-running operation, preventing other goroutines from gaining access to the resource in the meantime.

Figure 9.8 depicts a common scenario of where deadlocks commonly occur:

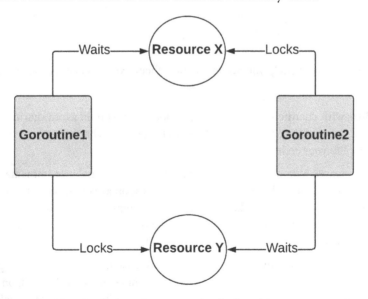

Figure 9.8 – A common deadlock problem

In this scenario, both goroutines need two resources to complete their work. Each of the goroutines is holding onto a resource while it is waiting for the second resource. Neither goroutine can complete its work and a deadlock occurs. This same scenario can also lead to a livelock, if each goroutine is polling to check the state of a resource or another goroutine. In this case, each goroutine is using CPU cycles but never finishes executing.

These four untestable conditions are actually a result of **poorly designed code** or a **faulty understanding of the behavior of concurrency mechanisms**. This is the main reason that we have started this chapter by thoroughly discussing and exploring the fundamentals and behavior of Go's concurrency mechanisms. These conditions can be detected by using good linters and a code review, but the best defense is to be aware of the problems when you are writing your code.

Figure 9.9 presents a summary of three rules of thumb when it comes to using concurrency:

Rule	Description
Share values with channels.	Don't write results to shared variables, use channels instead.
Defer lock releases.	Immediately defer the call to the unlock method of any acquired lock.
Wait for all children to complete.	The goroutine which starts other goroutines should wait for them to complete before shutting down.

Figure 9.9 – Concurrency rules of thumb

These three rules of thumb will help you minimize the difficult-to-detect concurrency issues we have discussed so far:

1. **Share values with channels**: As discussed previously, we should avoid sharing results using variables and pointers. Even when correctly protected by locks, channels are much more efficient and can simplify your code.

2. **Defer lock releases**: When using locks, you should get in the habit of immediately invoking the lock unlock/release with the `defer` keyword as soon as you acquire it. This will ensure that your function releases the lock, regardless of the completion logic branch or any potential errors. You should also consider whether you need a read-write lock, or whether you can avoid starvation by acquiring read mutexes, where possible.

3. **Wait for all children to complete**: As we have discussed, goroutines have a parent-child relationship with the goroutines they create. You should make use of synchronization mechanisms to ensure that the parent goroutine waits for all the goroutines it has created before shutting down to ensure that operations are completed correctly and resources are cleaned up correctly.

Testing alone **cannot prove the absence** of the four untestable conditions, but it can give us statistical confidence that these errors will not occur in production, for the scenarios that are important to our systems. Therefore, writing tests that cover the concurrent parts of our code is an important part of our testing strategy.

In the next section, we will look at how we can make use of the race detector to test the BookSwap application under concurrent conditions.

Use case – testing concurrency in the BookSwap application

The last section of this chapter is dedicated to the detection of concurrency issues in the BookSwap web application. We will make use of Go's race detector, alongside the testing strategies we have learned so far, to see what issues we can discover in the BookSwap application.

You might be wondering why we would worry about the concurrency aspects of the BookSwap application, since we have not used any locks, channels, or goroutines in the code base we have seen so far. Go's net/http library uses goroutines under the hood to serve HTTP requests, so the application can still have concurrency issues, even though it does not explicitly create its own goroutines and channels. This effect will be further amplified once the BookSwap application gets converted from a monolithic application to running in the microservice architecture we discussed in *Chapter 8, Testing Microservice Architectures*.

We already have all the tools available for writing tests that can simulate and verify the concurrent behavior of our application. We will write a test that creates BookSwap users concurrently:

```
func TestUpsertUser_Load(t *testing.T) {
if os.Getenv("LONG") == "" {
    t.Skip("Skipping TestUpsertUser_Load in short mode.")
  }
  userEndpoint := getTestEndpoint(t)
  requestBody, err := json.Marshal(map[string]string{
    "name":      "Concurrent Test User",
    "address":   "1 London Road",
    "post_code": "N1",
    "country":   "United Kingdom",
  })
  require.Nil(t, err)
  require.NotNil(t, requestBody)
  for i := 0; i < LOAD_AMOUNT; i++ {
    t.Run("concurrent upsert", func(t *testing.T) {
```

```
        t.Parallel()
        req := bytes.NewBuffer(requestBody)
        r, err := http.Post(userEndpoint, "application/json",
                            req)
        assert.Nil(t, err)
        body, err := io.ReadAll(r.Body)
        r.Body.Close()
        require.Nil(t, err)
        var resp handlers.Response
        err = json.Unmarshal(body, &resp)
        require.Nil(t, err)
        assert.Equal(t, http.StatusOK, r.StatusCode)
        assert.Nil(t, err)
        assert.NotNil(t, resp)
        assert.NotEmpty(t, resp.User.ID)
    })
  }
}
```

This benchmark test sends requests concurrently to the POST /users endpoint of the BookSwap application:

1. We declare a new test with the usual signature, which takes in a single *testing.T parameter. This test will run only when the LONG argument is passed to the test runner, as it required the BookSwap application to be up. The userEndpoint is returned by the getTestEndpoint helper function based on environment variables. For brevity, we have not included the implementation of this function here.

2. In the setup of the test, we marshal a map with the string key and value type, which contains all the JSON fields that we require to create a new user. We use the json.Marshal function in the standard library to do this. This function will return a slice of bytes, []byte, which we will use as the request body for our HTTP POST calls.

3. We repeat the same test in a for loop up to a LOAD_AMOUNT constant. The test runner will run the test in parallel according to the configuration that it has available. It's important to include this for loop; otherwise, our goroutine will only make a single call.

4. We set up the test to run in parallel using the t.Parallel method. Under the hood, this creates multiple goroutines and distributes the test iterations across them. This method takes in a function as a parameter, which will set up any local state and be run in each of the goroutines of the test.

5. In the loop, we convert the JSON byte slice into a buffer, which is required for the invocation of the `http.Post` function. This function takes in `usersEndpoint`, which will contain the URL to test against.

6. Once the HTTP request completes, it will return a response, against which we can make our assertions. We ensure that we close the body of the response in order to allow the same connection to be reused by another goroutine.

The simple configuration of the test will allow us to test our endpoint using a fixed number of concurrent requests. As we have seen in previous chapters, this function constructs the URL based on the environment variables specified for the application. If you want to run with the default values, set the `BOOKSWAP_BASE_URL` environment variable to http://localhost and the `BOOKSWAP_PORT` environment variable to `3000` to your terminal session.

By default, the number of goroutines that will be used by the benchmark is equal to the `GOMAXPROCS` variable. This variable will be equal to the number of CPUs on the machine it is running. Your OS decides what counts as a CPU, so for a four-core machine with hyperthreading, `GOMAXPROCS` will be `8`. If you would like to adjust the number of goroutines, you can easily configure this by changing this environment variable. Just as we have done with the `pprof` tool in *Chapter 8*, *Testing Microservice Architectures*, we run the `BookSwap` application with the race detector enabled:

```
$ go run -race cmd/main.go
```

The modification of resources and goroutines is happening in the `BookSwap` application itself, not the test code, so this is why we instrument the application and not the test code. Remember that the `BookSwap` application relies on a database now, we need to have PostgreSQL running and set the `BOOKSWAP_DB_URL` environment variable to the PostgreSQL connection string.

In a separate terminal window, we run the benchmark in the usual way, as running the benchmark in parallel is not visible outside the test configuration:

```
$ LONG=true go test  chapter09/user_create_test.go -v
```

We can then issue a `SIGINT` with *Ctrl + C* in the first console window in order to halt the race detector. If any data races are detected, they will be printed to the terminal, alongside the logs of `BookSwap`. Since the race detector is running separately from the test code, we cannot fail the test when data races are detected. We must therefore monitor the logs to see whether data races have been detected.

This simple technique can be used for end-to-end testing and integration testing of our application. You can use it to implement any user journey or sequence of requests. However, we should always bear in mind that the race detector is a limited tool and that no amount of testing can definitively and conclusively prove the absence of untestable concurrency issues.

Summary

In this chapter, we discussed the important and uniquely challenging topic of concurrency. As a good understanding of concurrency mechanisms is important to avoid issues, we started by learning the fundamentals of two of Go's most central concurrency mechanisms, goroutines and channels. Then, we looked at three applied concurrency examples, which helped us further explore their behavior.

Once we were familiar with how concurrency works, we started looking at some commonly occurring issues with concurrency. The Go race detector is a tool that can help us detect data races and provide pinpoint guidance to engineers to help them to resolve the issue. However, due to the importance of timing, it is not possible to conclusively prove the absence of concurrency issues, so careful design is always the first defense. Finally, we looked at an applied example of how to use benchmarks to make concurrent calls to the `BookSwap` application and detect issues with Go's race detector.

In *Chapter 10, Testing Edge Cases*, we will explore how to extend tests and ensure that our code is robust with Go's fuzzing capability.

Questions

1. What is the difference between concurrency and parallelism?
2. What operations do unbuffered channels support and how do they behave?
3. What is the difference between `sync.Mutex` and `sync.WaitGroup`?
4. What is a data race? What is a deadlock?
5. How do you use Go's race detector?

Further reading

- *Concurrency in Go: Tools and Techniques for Developers*, by Katherine Cox-Buday, published by O'Reilly
- *Network Programming with Go: Learn to Code Secure and Reliable Network Services from Scratch*, by Adam Woodbeck, published by No Starch Press
- *Introduction to Concurrency in Programming Languages*, by Matthew J. Sottile et al., published by Chapman & Hall

10
Testing Edge Cases

In the previous chapters, we discussed the implementation and testing of web applications. We made use of a variety of functional and non-functional tests to ensure that the individual services within our microservice architectures remained performant and delivered the correct functionality.

In *Chapter 4, Building Efficient Test Suites*, we discussed the definitions of edge and corner cases, as well as learning how to implement them using table-driven testing. For production systems, it would be nearly impossible to fully test complex systems, no matter how dedicated we might be to implementing tests across a variety of scenarios. Therefore, testing strategies should be designed with system requirements and user journeys in mind.

However, no matter how carefully we design and implement them, testing strategies also have their limitations. As discussed in *Chapter 9, Challenges of Testing Concurrent Code*, testing cannot prove the absence of concurrency bugs but does give us statistical confidence that these errors will not happen under the scenarios covered by our testing strategy. We also discussed how concurrency issues are not implementation bugs and are in fact system design faults, so a thorough understanding of Go's concurrency mechanisms is important in order to avoid such errors.

This chapter will explore the previously discussed topic of edge cases through a new lens. We will begin with an introduction to the concept of code robustness, which will allow us to write implementation code that is stable for a wide variety of inputs. Then, we will learn how to use Go's newly introduced fuzz testing capability to write tests that cover a wide variety of inputs. Finally, we will explore property-based testing, which allows us to write assertions for the output properties that really matter to us, as opposed to exact value matches. This is a different approach to test writing than we have seen so far and can make writing tests for edge cases much simpler.

In this chapter, we will cover the following topics:

- Definition and best practices for achieving code robustness

- Go's newly introduced fuzzing capability

- The concept of property-based testing and how we can implement it in Go

- How to use fuzzing and property-based testing in the `BookSwap` application

Technical requirements

You will need to have **Go version 1.19** or later installed to run the code samples in this chapter. The installation process is described in the official Go documentation at `https://go.dev/doc/install`.

The code examples included in this book are publicly available at `https://github.com/PacktPublishing/Test-Driven-Development-in-Go/chapter10`.

Code robustness

In *Chapter 4, Building Efficient Test Suites*, we discussed the types of variable values that our testing strategies should cover. Among these values, we identified three types of test cases covering our parameters:

- Base cases

- Edge cases

- Boundary cases

We further identified corner cases that occur when multiple input variables are supplied with edge case values. We should write test cases that cover a broad range of values for the inputs supplied to our functions.

In the world of microservice architectures, we often don't have control over which values are supplied to our services and functions, so the code we write should be stable under a variety of scenarios. In order to achieve this stability, we should implement a well-designed, well-tested robust code base.

Code robustness is an often overlooked quality that can help us achieve code that will remain stable even as it changes and goes through refactoring cycles. *Figure 10.1* presents the main characteristics of robust code:

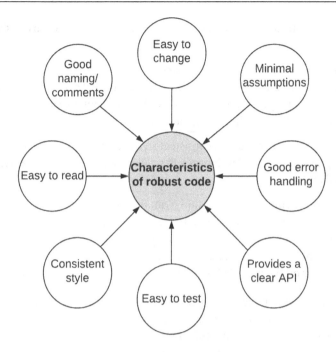

Figure 10.1 – Characteristics of robust code

In a nutshell, robust code is easy for developers to work on due to the following characteristics:

- **Easy to change**: Robust code is able to handle a wide variety of external factors, making it easier to refactor and change its dependencies. We will see that the other characteristics will contribute to making robust code easy to change.

- **Minimal assumptions**: Robust code makes minimal assumptions for the values of input parameters and resources that it has available. It should check any value that it does not generate internally to ensure that it is as expected.

- **Good error handling**: Robust code implements good error handling by checking for errors from external functions, gracefully ending current operations in the case of an error, and providing meaningful error messages to be returned to callers.

- **Provides a clear API**: Robust code provides an easy-to-read and comprehensible API for external callers. It should make it clear what parameter types it expects and what possible errors it can return. While documentation is a useful supplement, we should not rely on it. Robust code should use custom types and interfaces to leverage compiler checks and type safety as much as possible.

- **Easy to test**: Due to the characteristics we have already explored, robust code is easy to test. A clear API makes it easy for us to design our test cases, mock any dependencies required, or formulate any contracts for the implementation of contract testing. Good error handling, which returns well-formulated errors, allows us to write concise assertions that verify returned error messages.

- **Consistent style**: Robust code is written using a consistent set of practices, which includes naming standards and function structure. This can be achieved at code review as well as using Go's formatting and linting capabilities. You can read more about how to achieve this when using the `go vet` command in the official documentation – `https://pkg.go.dev/cmd/vet`.

- **Easy to read**: Robust code that uses a consistent style, handles errors, and provides good APIs reduces the cognitive load that developers will require to understand its behavior. This makes it easier to read and search through.

- **Good naming/comments**: Documentation is another important but often overlooked aspect of writing code. Variable naming should be short, but representative of the functionality it provides. Functions and types should have accompanying documentation that clearly states their expected behavior and the functionality it provides. You can read more about how to write Go comments on the official blog – `https://tip.golang.org/doc/comment`.

Designing code with robustness in mind will help us create stable systems and microservice architectures. One of the common ways that developers implement systems is by emulating the principles of the **Unix philosophy**, which has been established by its creators and the community. It states that robustness results from transparency and simplicity, which are principles that relate well to Go software development. Looking at these principles, we can see that they are reflected in the characteristics we have examined in *Figure 10.1*:

- Transparent code is easy to read and understand. Readability is aided by consistent styling, good naming/comments, and clear API definitions.

- Simple code is also easy to understand, but it is also uncomplicated and provides well-defined functionality that can be easily reused. This provides our systems with modularity, which allows us to reuse them to solve a wide variety of problems and situations. However, it should also be able to gracefully handle error cases or situations that it was not designed to accommodate.

Just like Linux, open source software and libraries are generally seen as more robust than their proprietary counterparts, as they have a wider audience that can find and correct errors. This is one of the reasons that we have only explored open source libraries and tools throughout our entire exploration of TDD.

The opposite of robust code is fragile and error-prone code. This type of code is complicated to understand and can be difficult to refactor, even when using the strategies we explored in *Chapter 7, Refactoring in Go*. Often, code refactoring involves making code changes to add robustness to code using some of the best practices we will explore in the next section.

Best practices

Now that we have a good grasp of the characteristics of robust code, we can begin to look at some best practices for implementing it in Go. We can begin our exploration by looking at a simple example of a fragile piece of code and exploring what we can do to make it more robust.

We will write a function that returns the values contained inside a Go map in key-sorted order. To achieve this, our function will do the following:

1. Extract the keys contained inside the map.

2. Sort the keys according to a given order parameter.

3. Extract the values corresponding to their keys and return them as a slice.

> **Map refresher**
>
> Go maps are dynamic, unordered collections of key-value pairs. Values can be accessed and modified using a unique key. They are represented using the built-in map type. The zero value of a map is nil, so it is initialized using a make function. Since Go 1.12, the fmt package will print maps in key-sorted order, but it is important to remember that the collection is unordered.

Based on this knowledge, we can write a simple function to return the sorted values from a map:

```go
var input map[int]string
func GetValues(dir string) []string {
  var keys []int
  for k := range input {
    keys = append(keys, k)
  }
  if dir == "asc" {
    sort.Ints(keys)
  }
  if dir == "desc" {
    sort.Slice(keys, func(i, j int) bool {
      return keys[i] > keys[j]
    })
  }
  var vals []string
  for _, k := range keys {
    vals = append(vals, input[k])
```

```
    }
    return vals
}
```

This example function does work, returning sorted values, but it does have some areas for improvement to make it less fragile:

1. **Global variables**: The input map is a global variable, defined outside the scope of the function. The dependency between the function and the map is unclear from looking at the signature.

2. **Function name and signature**: The function name, `GetValues`, does not give any indication of the sorting functionality. It is also unclear what the `dir` parameter is used for and what its allowed values are.

3. **Nil input behavior**: As previously mentioned, the zero value of a map is nil. The function immediately uses the input map, without checking for nil. The `range` operator is able to handle a nil map passed to it without panic, but it is unclear what the expected behavior of the `GetValues` function is in this case.

4. **Input validation**: The function allows two values for sorting the keys in ascending or descending order, as specified by the `dir` parameter. The function handles the value of the parameter and performs the corresponding sorting, but the values will simply remain unsorted in the case of another value. Instead, it would be better if the caller of the function received some indication that the function was not able to perform its intended work.

5. **Hardcoded strings**: Another issue with the `dir` parameter is that it is validated against hardcoded values defined inside the function. Unless the value of the sorting direction is the same, including the letter case, the function will not match it. Furthermore, the caller has no idea what the accepted values are unless the implementation code is inspected.

6. **Inconsistent style**: The function uses the built-in `sort` package to sort the keys in the correct order. This is a lot better than implementing a sorting algorithm from scratch, but the two sorting cases are implemented using different `sort` package functions.

7. **Memory allocation**: The slices used for the keys and sorted values are declared as their zero values, even though we already know how many keys are values we will be sorting. Under the hood, the Go runtime will have to expand the underlying arrays and copy the data as the number of values grows.

We can add robustness to this function by addressing the issues we've found with the original implementation. This code refactoring is relatively straightforward, as this function is currently operating in isolation from any other code. A revised version is as follows:

```
type SortDirection int
const (
    ASC SortDirection = iota
```

```
   DESC
)
// GetSortedValues returns the key-sorted values of a given
map.
func GetSortedValues(input map[int]string, dir SortDirection)
([]string, error) {
  if input == nil {
     return nil, fmt.Errorf("cannot sort nil input map")
  }
  keys := make([]int, 0, len(input))
  for k := range input {
    keys = append(keys, k)
  }
  switch dir {
    case ASC:
      sort.Slice(keys, func(i, j int) bool {
        return keys[i] < keys[j]
      })
    case DESC:
      sort.Slice(keys, func(i, j int) bool {
return keys[i] > keys[j]
})
    default:
      return nil, fmt.Errorf("sort direction not recognized")
    }
    vals := make([]string, 0, len(input))
    for _, k := range keys {
      vals = append(vals, input[k])
    }
    return vals, nil
}
```

The changes we have made to this simple function have addressed a lot of the issues that we previously identified:

1.   We introduce a new SortDirection type to replace the string value of the dir parameter. This type is used to create an enum with the acceptable sort direction values.

2. The signature of the function has been changed to take the input map as a parameter, removing its dependency from the input map global variable. The function also returns a second error value in the case that it cannot complete its operation.

3. Alternatively, we could have allowed the `GetSortedValues` function to take a sorting function as a parameter, allowing the caller to implement their custom `sort` functions. This would allow us to move the entire sorting logic outside the function, but would also give calling code a lot more flexibility.

4. In the case that the input map is nil, the function returns an error and stops being executed. This is the behavior we have decided to implement for the function, making it clear to the user that an uninitialized map is considered an error case.

5. Slices used for saving keys and values are initialized with the same capacity as the length of the input map. Appending values to these slices will not cause data reallocations and copying under the hood. As this is a small sample function, we will assume that the size of the map is enough to load in memory, but this might not be the case when processing very large datasets.

6. We make use of a `switch` statement to check the values of the newly implemented `SortDirection`. The function has two acceptable cases for ascending and descending sort orders and returns an error in the case that another `enum` value is introduced without the correct implementation. Both cases implement sorting using the `sort.Slice` function.

These changes have added robustness to our simple function, making it easier and more intuitive to use. We have also seen that adding code robustness is all about small changes, which add up to big improvements in our code stability and readability.

Once we get into the habit of writing code with robustness and stability in mind, it becomes a habit to build it into our solutions, removing the need to return and refactor it later. As we have stated multiple times, test code is just as important as implementation code. Therefore, it should also be designed with robustness in mind. In the next sections, we will explore two strategies for writing robust test code that verifies edge cases.

Usages of fuzzing

As discussed previously, it is very difficult to write tests that cover all possible user scenarios and parameter value ranges. The number of test cases to write and maintain can become even more time-consuming than project work. In this section, we will explore Go's fuzz testing capability, which can help us write tests that cover a wide variety of inputs.

Fuzz testing is a powerful technique that has been used to find bugs in a wide variety of software systems, including the Go standard library itself. It involves generating a wide variety of values and using them as input to the UUT. The generated values stress-test the UUT and help uncover bugs or unexpected behavior such as panics, memory leaks, or incorrect outputs.

Fuzz tests are automated, black-box tests that can be used to detect any potential functional or security issues in our system. They are usually run using a **fuzz tool**, which takes care of value generation, test execution, and error detection. In this section, we will focus on using fuzzing to detect functional errors.

Figure 10.2 presents an overview of the steps involved in fuzz testing:

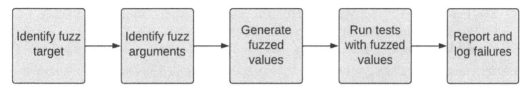

Figure 10.2 – The main steps of fuzz testing

Writing fuzz tests is not very different from regular unit tests:

1. **Identify the fuzz target**: Just like with normal tests, we begin by identifying the UUT. The fuzz target will be the function that we will cover in our tests.

2. **Identify fuzz arguments**: The fuzz target is only suitable for fuzz testing if it takes in at least one parameter. These are the parameters that will be generated by the fuzz tool and used as input for the previously identified fuzz target.

3. **Generate fuzzed values**: Once the test is specified, the fuzz tool will begin to generate the random values for our fuzz arguments.

4. **Run tests with fuzzed values**: The tests are executed with the generated fuzzed values. Typically, fuzz tests are fast-running unit tests, as they will be run with a large amount of generated fuzzed values.

5. **Report and log failures**: The test runner will execute the tests, logging and reporting failures. Just like unit tests, fuzz tests can include assertions and verifications.

Fuzz testing can be used to validate any input values that the UUT or fuzz target has not generated. These can be values coming from other services in the system or from outside sources. It can be applied to files, policies, applications, and libraries.

Fuzz testing in Go

The ability to fuzz-test is an important component of a well-rounded testing strategy. Fuzzing was added to the standard Go testing library in Go version 1.18. This feature was highly anticipated by the Go community, making writing fuzzed tests as easy as writing unit tests.

Just like tests and benchmarks, fuzzed tests must follow a few rules:

1. Tests must begin with the `Fuzz` prefix. We notice that tests are exported functions, defined by starting with a capital letter. For example, a fuzz test for our `GetSortedValues` function would be named `FuzzGetSortedValues`.

2. Tests must be defined in test files named with the `_test.go` postfix. As with other test files, we should use the name of the source file to name our test file. For example, if our sorting function was defined in a `sort.go` file, then its corresponding test file could be `sort_test.go`.

3. Tests must accept a single `*testing.F` parameter and have no return values. This is the way fuzzed tests interact with the test runner and fuzzing tool.

4. The fuzz target is defined by calling the `Fuzz` function on the `*testing.F` parameter. This function takes in a `*testing.T` parameter, followed by the fuzzing arguments. There can only be one fuzz target per test and the calls on the UUT will happen inside the fuzz target.

5. Fuzz arguments are added to the fuzzing tool using the `Add` function on the `*testing.F` parameter. This will instruct the tool to generate values to be used in the fuzz target.

6. Fuzzing arguments can be of the following types:

    I.   `string, []byte`

    II.   All `int` types, including `rune`

    III. All `uint` types, including `byte`

    IV.   All `float` types

    V.   `bool`

7. Due to a large number of test runs, fuzz tests will be run in parallel. They should therefore be deterministic.

8. Fuzz tests are run alongside your other unit tests using the `go test` command or with the `-fuzz` flag followed by a test name or package. Again, this is similar to how we run benchmark tests.

The fuzzing tool monitors the test runs and reports errors that have occurred. Fuzz tests can fail for a few reasons: panics, test failures, non-recoverable errors, or timeouts. By default, the timeout for a fuzz target is 1 second, so your tests should be fast.

Fuzzed tests will continue to run until a failing input is found or until the user cancels the test run manually. Alternatively, we can supply a maximum execution time or the maximum number of iterations using the `-fuzztime` command-line parameter.

A simple fuzz test

We can write a simple fuzz test for the GetSortedValues function we wrote in the previous section:

```
func FuzzGetSortedValues_ASC(f *testing.F) {
  input := map[int]string{
    99: "B",
    0:  "A",
  }
  f.Add(3, "W")
  f.Fuzz(func(t *testing.T, k int, v string) {
    input[k] = v
    keys := make([]int, 0, len(input))
    for k := range input {
      keys = append(keys, k)
    }
    sort.Ints(keys)
    sortedValues, err := GetSortedValues(input, ASC)
    require.Nil(t, err)
    require.NotNil(t, sortedValues)
    for index, v := range sortedValues {
      key := keys[index]
      assert.Equal(t, input[key], v)
    }
  })
}
```

We have written the fuzzed test according to the rules we have discussed and with the same techniques we use for unit tests:

1. We declare a fuzz test using the required signature. The test starts with the Fuzz prefix and takes in the *testing.F parameter.

2. We add two fuzz testing arguments to the fuzz testing tool using the f.Add method, one for the map key of the int type and one of the string type for the map value. These values will be generated by the fuzzing tool. Both of these types are accepted for fuzzing arguments.

3. We define the fuzz target using the f.Fuzz method. This method takes a function as a parameter, which itself takes the fuzzing arguments as parameters. The function also takes in a *testing.T parameter, which makes it possible for us to write test assertions inside the fuzz target.

4. Inside the fuzz target, we write the testing code. We add the fuzzing arguments to the map, using the generated values to test our functionality. Then, we extract the keys from the map and sort them in ascending order.

5. We invoke the `GetSortedValues` function, which is our UUT for this test, passing it the input map that now contains the fuzzing arguments.

6. At the end of the test, we use the previously sorted slice of keys to assert that the values returned are sorted correctly.

We have successfully written our first fuzzed test. We can run it using the `go test` command with two configuration flags:

```
$ go test -fuzz FuzzGetSortedValues_ASC -fuzztime 5s ./
chapter10/fragile-revised -v
=== FUZZ  FuzzGetSortedValues_ASC
fuzz: elapsed: 0s, gathering baseline coverage: 0/711 completed
fuzz: elapsed: 1s, gathering baseline coverage: 711/711
completed, now fuzzing with 8 workers
fuzz: elapsed: 3s, execs: 10707 (3569/sec), new interesting: 54
(total: 765)
fuzz: elapsed: 5s, execs: 17519 (3262/sec), new interesting: 67
(total: 778)
--- PASS: FuzzGetSortedValues_ASC (5.14s)
PASS
ok      github.com/PacktPublishing/Test-Driven-Development-
in-Go/chapter10/fragile-revised      5.303s
```

The `-fuzz` flag instructs the test runner to execute the fuzzed test specified by name, while the `-fuzztime` flag specifies that the test should run for a maximum of 5 seconds. The output of our test run highlights some key metrics of the test run:

- `elapsed` indicates the amount of processing time
- `baseline coverage` indicates the number of scenarios that are applied to measure the coverage provided by the tests
- `execs` indicates the number of test cases that have been run with the fuzz target
- `new interesting` is the number of new inputs that are identified that expand the coverage of the fuzzed test

Now that we understand how to write and run fuzzed tests, we are ready to add them to our own testing strategies. However, it does have some drawbacks. *Figure 10.3* presents some of the advantages and disadvantages of fuzz testing:

Advantages	Disadvantages
+ Easy to use and implement	- Does not replace traditional testing
+ Can be used early in the development lifecycle	- Does not provide guarantees
+ Detects a wide variety of bugs	- Memory & CPU intensive

Figure 10.3 – Advantages and disadvantages of fuzz testing

The advantages of fuzz testing are as follows:

1. **Easy to use and implement**: As fuzzing integrates with Go's `testing` package, we can easily write fuzzed tests for anything in Go. However, it's important to keep the scope of the tests small so that they can be executed quickly and efficiently.

2. **Can be used early in the development life cycle**: As we have seen in our simple example, fuzzed tests can be written for functions or even small units of code. This makes it easy to leverage them at any stage of the development life cycle.

3. **Detects a wide variety of bugs**: Fuzzed testing generates values that cover edge cases and run over many executions. This makes it a great tool for detecting bugs that would otherwise not have been possible to find.

The disadvantages of fuzz testing are as follows:

1. **Does not replace traditional testing**: Fuzzing complements rather than substitutes the types of tests that we have explored throughout this book. Therefore, it can take additional engineering effort to write these tests.

2. **Does not provide guarantees**: Fuzzed tests only provide an indication of the stability of the UUT, not a guarantee. As it generates random values, it can only indicate to developers the presence of bugs for the inputs it does cover.

3. **Memory- & CPU-intensive**: As we have seen from our example output, fuzzed tests are run in parallel over a large number of scenarios. This makes them more memory- and CPU-intensive than unit tests.

Despite its drawbacks, fuzz testing is a useful and powerful testing technique that complements all the traditional testing methods we have explored so far. As it is able to generate a wide variety of inputs, fuzz testing is also an important tool that can help uncover security vulnerabilities. In security fuzz testing, we input malicious data to a program, while in functional fuzz testing, we input invalid data. We will not focus on security testing in this book, but it is another great use for fuzzing. It is especially useful for ensuring our systems remain stable in edge cases or when processing user inputs.

Property-based testing

Fuzz testing is a great step forward when testing edge cases in our application. We can think of it as analogous to chaos testing, where we test a huge variety of edge cases in the hope of detecting an error. However, we do not have any control over the random inputs. This leads to two problems:

- We test a large number of irrelevant scenarios that are unlikely to happen in our system.

- We don't know whether the scenarios that really matter have been covered by our fuzzed tests. Instead, it would be great if we had a more structured approach available to us.

Property-based testing is a testing technique that involves testing a program against a set of properties or specifications that are important to our user journeys and system behavior. This allows engineers to follow a systemic approach to testing, as opposed to focusing on verifying inputs.

In property-based testing, we generate random inputs that satisfy the set of constraints or properties that we have identified. The generation aspect ensures that we test against a larger space of edge cases than would have been possible with traditional, manually written tests. The focus on properties ensures that we cover the edge cases that matter to our applications. Again, this does not guarantee the absence of bugs, but it does ensure that we spend our time testing the things that matter.

The `testing/quick` package offers testing helper functionality that we can leverage to implement property-based tests:

- The `quick.Check` function takes in a function with a `bool` return value and searches for arbitrary values that make the input function return `false`.

- The `quick.CheckEqual` function takes in two functions and looks for an input for which the functions return different results.

- The `quick.Generator` interface defines a Generate method that custom types can implement. Once they satisfy this interface, we can generate random values for our custom types using the `quick.Value` function. This gives us the flexibility to generate values for any exported type.

The Check functions of the `quick` package also take in a `*quick.Config` parameter that allows us to configure our test run with the maximum number of iterations or another random generator.

This type of testing is intuitive and easy to achieve. Looking back at the fuzzing example we implemented, the verifications in the fuzz target only asserted the order of the elements, not the values themselves. In fact, without realizing it, we wrote our first property-based test. The real value of property-based testing, however, lies in its search for failing function inputs, as opposed to generating fully random values.

We can re-implement our previously implemented fuzz test with property-based testing in mind:

```go
func TestGetSortedValues_ASC(t *testing.T) {
    input := map[int]string{
```

```
      99: "B",
      0:  "A",
    }
  isSorted := func(k int, val string) bool {
    input[k] = val
    keys := make([]int, 0, len(input))
    for k := range input {
      keys = append(keys, k)
    }
    sort.Ints(keys)
    sortedValues, err := fr.GetSortedValues(input, fr.ASC)
    if err != nil || sortedValues == nil {
      return false
    }
    for index, v := range sortedValues {
      key := keys[index]
      if input[key] != v {
        return false
      }
    }
    return true
  }
  if err := quick.Check(isSorted, nil); err != nil {
    t.Error(err)
  }
}
```

The structure of the test is different, but it uses the same verifications as in the fuzzing test:

1. The test uses the regular unit test signature, starting with the Test prefix and taking in a single parameter of *testing.T.

2. Inside the test, we declare an isSorted helper function, which takes in the two arguments we will generate, one for the key and one for the value of our new map entry. It also returns a bool value, making it suitable to be used with the quick.Check function.

3. Inside the function, we add the generated values to the input map. Then, we copy the keys and sort them. We call the GetSortedValues function and our UUT and get the actual values to verify.

4. In the case of an error or a nil slice, we return `false`, stopping the test. We also return `false` if the sorted values are not as expected. This will signal to the `quick.Check` function that an error has occurred.

5. Inside the test, we pass the `isSorted` helper function to the `quick.Check` function and fail the test if it returns an error.

In the case of error, the `quick.Check` function will report the values that have caused the failure. Forcing the test to fail, we will receive an output with information about the inputs that caused the failure:

```
$ go test -run TestGetSortedValues_ASC ./chapter10/fragile-
revised -v
=== RUN    TestGetSortedValues_ASC
  sort_test.go:62: #1: failed on input 73546389, "\U000773b8"
--- FAIL: TestGetSortedValues_ASC (0.00s)
FAIL
exit status 1
FAIL    github.com/PacktPublishing/Test-Driven-Development-
in-Go/chapter10/fragile-revised       0.154s
```

The input values causing the failure can be used by engineers to debug the application and fix the cause of the failure.

The two testing techniques we have covered in this chapter, fuzz testing and property-based testing, allow us to take advantage of value generation and test a wide variety of edge cases for our system inputs. These testing techniques are complementary to the robust code best practices discussed at the beginning of the chapter and allow us to ensure the stability and reliability of our services.

Use case – edge cases of the BookSwap application

This chapter has taught us two new testing techniques: fuzzed testing and property-based testing. We have learned how to apply these to a simple function that provided the functionality of returning the key-sorted values contained inside an input map. In this section, we will end our exploration with a discussion of how these techniques can be applied to the BookSwap application we have built so far.

As previously discussed, robust code should test any variables that it does not generate itself. We named these inputs as being from untrusted sources. *Figure 10.4* presents all the inputs that can be considered untrusted from the viewpoint of the processing service:

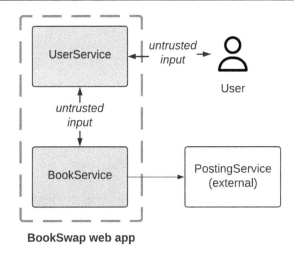

BookSwap web app

Figure 10.4 – Untrusted input in the BookSwap application

We identify two sources of untrusted input within the services of BookSwap:

- UserService receives input from the user. As we have no control over what the user submits, this interaction is a source of untrusted input.

- UserService receives information from BookService. As you may remember from our discussion in *Chapter 8*, *Testing Microservice Architectures*, microservices change without any central oversight. Therefore, as UserService does not have any control over the functionality provided by this external service, this interaction is a source of untrusted input.

- The same applies to BookService, which receives requests from UserService. While this interaction does not actively send input to BookService, the request still contains untrusted information.

Based on these insights, we can identify the need to implement fuzz testing for the HTTP handlers exposed by our web application. We will test the user creation operation that we have also tested in previous chapters. Fuzz testing of HTTP endpoints can easily achieve this using Go's fuzzing capability:

```
func FuzzTestUserCreation(f *testing.F) {
  // other initialization
  f.Add("test user", "1 London Road", "N1", "UK")
  f.Fuzz(func(t *testing.T, name string, address string,
    postCode string, country string) {
    requestBody, err := json.Marshal(map[string]string{
      "name":     name,
      "address":  address,
```

```
        "post_code": postCode,
        "country":    country,
    })
    require.Nil(t, err)
    req := bytes.NewBuffer(requestBody)
    resp, err := http.Post(userEndpoint, "application/json",
req)
    assert.Nil(t, err)
    defer resp.Body.Close()
    assert.Equal(t, http.StatusOK, resp.StatusCode)
    assert.Nil(t, err)
    assert.NotNil(t, resp)
    })
}
```

Other than the interaction with the fuzz testing tool for setting up the fuzzing target and arguments, the body of this test is identical to the HTTP testing we are already familiar with. We add four fuzzing arguments for the fields that the user creation request needs. Inside the fuzzing target, we create and marshal the request to JSON format. Finally, we send the request to the /users endpoint using the http.Post function. Just as the previous fuzz testing example we have seen in this chapter, we run this test using the LONG=true go test -fuzz FuzzTestUserCreation -fuzztime=5s ./chapter10 -v command. It requires the BookSwap application to be up and running, which we can easily do with the Docker command we've seen so far docker compose -f docker-compose.book-swap.chapter10.yml up --build. The test also relies on environment variables to construct the URL under test. If you want to run with the default values, set the BOOKSWAP_BASE_URL environment variable to http://localhost and the BOOKSWAP_PORT environment variable to 3000 to your terminal session.

This brings us to the end of our exploration of fuzz testing and property-based testing, which are two related techniques that allow us to easily write testing strategies that cover a wide variety of edge cases. Both of these techniques integrate well with Go's testing package and can be easily implemented at any level of our application testing.

Summary

In this chapter, we revisited our understanding and approach to covering edge cases of input variables. We began our discussion with the concept of code robustness, focusing on writing code that can handle unexpected inputs and errors. Once code robustness becomes part of our mindset, we start to understand the edge cases of our code. However, it can be very difficult to manually implement tests to cover all these values.

Two testing techniques leverage input generation to make this task easier: fuzz testing and property-based testing. Both of these techniques can be implemented with Go's testing library, allowing us to easily leverage the benefits of these techniques to verify a broad range of edge cases to our components at any level. Finally, we looked at an applied example of how to use fuzz testing together with the BookSwap application's user creation flow, ensuring that it is stable for generated values.

In *Chapter 11, Working with Generics*, we will explore another newly introduced Go feature, namely generics. We will see how it can make our implementation and testing code simpler and easier to use.

Questions

1. What are some of the characteristics of robust code?
2. What is fuzz testing?
3. What is the signature of a fuzzed test in Go?
4. What is property-based testing?
5. What are the untrusted inputs of the UUT?

Further reading

- *The Art of Clean Code: Best Practices to Eliminate Complexity and Simplify Your Life*, Christian Mayer, published by No Starch Press
- *How to Break Software: A Practical Guide to Testing*, James Whittaker, published by Pearson

11

Working with Generics

The previous chapters have covered all the aspects of testing and code design, gradually increasing in complexity. These tests will allow us to verify the behavior of a wide variety of applications, including microservices architectures and services that persist data to a database service. We have all the knowledge to perform the functional and non-functional testing of a variety of systems.

We have also discussed how systems need to change and evolve, as engineering organizations grow and new functionality is added. In *Chapter 9, Challenges of Testing Concurrent Code*, we discussed how to make the most of computing resources by making use of Go's concurrency mechanisms, enabling us to serve production traffic. In *Chapter 10, Testing Edge Cases*, we explored two testing techniques to cover system edge cases, including Go's fuzz testing capabilities.

Go's **1.18** release introduced support for generics, alongside adding fuzz testing to the `go test` toolset. Generics was a major, much-anticipated feature by the Go community. It allows us to write functions, types, and data structures that can work with any type, rather than being limited to one type when we declare them. A newly released language feature is another reason why we might change our code, alongside the maintenance and feature development processes we have discussed.

This chapter will explore the implementation and testing of generic code. We will begin by exploring how to write generic code, exploring both use cases and syntax. Then, we will explore how to test generic code, including where generics can be used to streamline our test code. Finally, we will reflect on all we have learned and summarize some of the testing best practices that we have explored throughout this book.

In this chapter, we will cover the following topics:

- The basics of writing generic code in Go
- Testing generic code
- Leveraging generics for test utilities
- A summary of the testing best practices we have covered in this book

Technical requirements

You will need to have **Go version 1.19** or later installed to run the code samples in this chapter. The installation process is described in the official Go documentation at `https://go.dev/doc/install`.

The code examples included in this book are publicly available at `https://github.com/PacktPublishing/Test-Driven-Development-in-Go/chapter11`.

Writing generic code in Go

The introduction of generics in Go was a highly debated and anticipated feature. Some developers felt that introducing it would contradict Go's core principles of simplicity, while others felt like it was a sign of maturity and would allow them to write better production code. As with every technical solution or design decision, there is a trade-off between the advantages and disadvantages.

As previously mentioned, **generics** refers to the ability to write code that works with different data types without being limited to a specific type. In the absence of generics, we have used Go interfaces to implement generic behavior in Go. In *Chapter 3*, *Mocking and Assertion Frameworks*, we explored the power of interfaces and have seen how they can be used for wrapping and replacing dependencies. While interfaces are not the same as generics, they provide a way to achieve similar goals of **flexibility** and **code reuse**.

Figure 11.1 provides a comparison of generics and interfaces:

Generics	Interfaces
+ Specify type	+ Specify behavior
+ Built into the language	+ Defined by the application
+ Limited scope	+ Wide scope

Figure 11.1 – Comparison of generics and interfaces

While both generics and interfaces provide code flexibility and polymorphism to our code, they have the following main differences:

1. Generics are a way to specify the type, while interfaces specify behavior. As we have seen, interfaces are collections of methods that must be defined by structs in order to satisfy them. On the other hand, generics give us the ability to specify the types of parameters that can be used.

2. Generics are built into the language, while interfaces are defined by the application. Interfaces are defined by engineers as part of their code bases, which makes it easier to define them to include any behavior required by the application. The specifications of generics are built into the language and can be shared across code bases.

3. Generics have limited scope, while interfaces have a wide scope. As they are built into the language, they are simple enough to implement solutions to a wide variety of problems. On the other hand, interfaces are expressive and can define complex behavior.

Both generics and interfaces are implemented by the Go compiler, so they are both statically type-checked. However, interfaces ensure that specific methods are available for a parameter, while generics do not provide these guarantees.

Generics in Go

The ability to write generic code is a core part of other strongly typed programming languages such as Java, C#, and C++. Its addition to Go gives us the ability to write more flexible and reusable code. Let us look at some examples of how we can leverage this new ability.

There are three main components for specifying generic code:

1. **Type parameters** are the placeholder type specifications that will be used with generic code. They are typically denoted with one single letter, for example, T, and allow us to reference the placeholder type in our implementation. A generic function or type is defined by this placeholder as part of its specification.

2. **Type constraints** help us define rules or subtypes for the type parameters. Constraints are not full specifications like interfaces, but they allow us to restrict type parameters to certain properties.

3. **Type arguments** are the type that is passed to the generic function or type upon invocation, which specifies the type of data we will be using. Type arguments are used in place of the type parameter placeholders declared by the function or type signature.

4. **Type inference** is the process that takes place under the hood to determine the type of a variable, without its explicit type specification. This also allows us to write less verbose generic code.

These four components work together to make it easy for us to write generic code. *Figure 11.2* demonstrates each component with a simple example of a generic function:

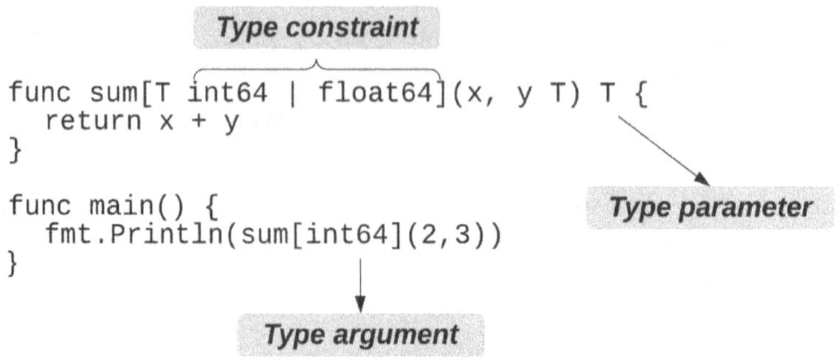

```
                    Type constraint
                    ⎴
func sum[T int64 | float64](x, y T) T {
    return x + y
}

func main() {                          Type parameter
    fmt.Println(sum[int64](2,3))
}
              │
              ↓
         Type argument
```

Figure 11.2 – Generics example on a simple function

The example defines a generic sum function that takes in two parameters. It defines a type parameter, T, which must satisfy the specified type constraints. In the main function, we invoke the function with an int64 type argument and two parameters. The compiler uses type inference to ensure that the passed parameters comply with the type constraint and replace the type parameter with the type argument supplied in the main function. We will explore how to write and test generic Go code in the next sections.

Exploring type constraints

Type constraints are a central component of generics, as they allow us to limit the data types that our generic code can be used with. This makes it easier for us to write safe and simple code that operates on types that meet specified conditions.

Type parameters are declared inside square brackets ([]) in Go. The type parameter defines a name and possible constraints for a type. We can declare constraints in a few ways:

- The any interface is an alias for the empty interface, interface{}, and allows all types to be used. This allows us to build functions and types without constraints. For example, we can make the sum function accept any parameter type by using the any interface:

```
func sum[T any] (x, y T) T {
    // implementation
}
```

- The comparable interface is a pre-declared type constraint that denotes types that can be compared using the == operator. These types are string, bool, all numbers, and composite types containing comparable fields. For example, the sum function will now allow these types as parameters, but we will need to implement the sum logic accordingly, as the + operator will no longer be implemented by all these types:

```go
func sum[T comparable](x, y T) T {
    // implementation
}
```

- **Type sets** can be created using the | operator. This allows us to create constraints that contain multiple types without having to wrap them in a custom interface. This is the example we have seen in *Figure 11.2*, where the sum function allows the int64 and float64 types, which already support the + operator:

```go
func sum[T int64 | float64](x, y T) T {
    // implementation
}
```

- **Custom type constraints** can also be created as an interface and reused in our code. These are also declared using the | operator. For example, the Number interface allows the int64 and float64 types, which we can then use in the specification of the sum function, resulting in the same specification as the previous type sets:

```go
type Number interface {
    int64 | float64
}
func sum[T Number](x, y T) T {
    // implementation
}
```

- The ~ keyword can be used to restrict all custom types that have the same underlying type. This allows us to encompass custom types into our constraints. For example, the Number interface will now allow any int- and float64-based types:

```go
type Number interface {
    ~int64 | ~float64
}
```

- The constraints package defines some useful constraints that can be used together with your generic code. This package contains numerical and ordered constraints that you might find useful. For example, we can modify the sum function to accept all signed integers by using this package:

```go
func sum[T constraints.Signed](x, y T) T {
    // implementation
}
```

Generics can be applied to functions and structs, allowing us to create generic data structures and reusable behaviors. This can streamline our code and remove the need to define code multiple times in order to accommodate different types of underlying data. In *Chapter 10*, *Testing Edge Cases*, we implemented a function that sorts map values based on key values:

```
// GetSortedValues returns the key-sorted values of a given
map.
func GetSortedValues(input map[int]string, dir SortDirection)
([]string, error) {
    // implementation
}
```

This very useful function has a major limitation in that it only works for a single type of map with an int key and a string value – map[int]string. This limitation makes it difficult to reuse this code for other map types.

In the days before Go generics, we would have to create a function for each type of map:

```
 // GetSortedFloatValues returns the key-sorted values of a
given map with int key and float64 values.
func GetSortedFloatValues(input map[int]float64, dir
SortDirection) ([]string, error) {
    // implementation
}
```

Since Go does not allow function overriding, we also need to use different names for the function taking different parameters. This can make it difficult to create libraries, even if we extract helper functions to help us reuse the implementation code inside the function.

This is much simplified by generics, which allow us to parameterize the key and value types of the sort function:

```
// GetSortedValues returns the key-sorted values of a given
map.
func GetSortedValues[K ~int, V comparable](input map[K]V, dir
SortDirection) ([]V, error) {
    // implementation
}
```

The signature of the function has been changed to leverage the power of generics to create a reusable function:

- The function has two type parameters, one for the key type and one for the value type. The key value is restricted to any `int` type using the ~ keyword, while the value is opened to any type that complies using the `comparable` interface.

- The `map` parameter is declared using the key and value placeholder types, `map [K] V`. Only parameters that comply with the type constraints will be accepted by the function.

- The return value of the function is adjusted to return a slice of value types, `[] V`. This ensures that values returned are of the type as passed in by the input map.

Once we open up our functions to receive generic types, it is also important to apply meaningful constraints to ensure that our code continues to provide useful functionality. The function signature also serves as a user manual for callers, so we should continue to guide them, even though we are adding flexibility to our code.

Table-driven testing revisited

Now that we understand the basics of implementing generic code, we can turn our attention to testing it. The adoption of generics throughout the Go community is still in its beginning stages. We have already established that generic code is statically enforced by the compiler, but does this increase in flexibility lead to more complex test code?

We can continue our exploration using the generic `GetSortedValues` function that we have implemented. Tests should now be written to assert the behavior of the function for a variety of input types and values. We can achieve this by using the table-driven testing technique that we explored in *Chapter 4, Building Efficient Test Suites*. The implementation of generic table-driven testing follows a series of steps.

Step 1 – defining generic test cases

We begin by creating a generic test case type to save our input map and values:

```
type testCase[K ~int, V comparable] struct {
  input  map[K]V
}
```

A generic test case allows us to construct an input map with the correct key and value type constraints, which are the same as we have used for the generic `GetSortedValues` function. We will use the input map as a parameter for the UUT.

Step 2 – creating test cases

As we remember from our introduction to table-driven testing, we typically construct a map of test cases with the scenarios that we will be running our tests over. Due to the differences in type, we construct multiple test case maps:

```
type CustomI int
testStrings := map[string]testCase[int, string]{
    "unordered":        {input: map[int]string{99: "A", 50:"X"}},
    "empty map":        {input: map[int]string{}},
    "negative values": {input: map[int]string{-99: "A", -1:"X"}},
}
testFloats := map[string]testCase[CustomI, float64]{
    "unordered":       {input: map[CustomI]float64{99: 1.2, 0:
4.6}},
    "empty map":       {input: map[CustomI]float64{}},
    "negative keys": {input: map[CustomI]float64{-99: 1.2, 0:
4.6}},
}
```

In this example, we create two maps with three scenarios each. The first `testCase` map has an `int` key type and a `string` value type. The second `testCase` map has an `int`-based custom type, `CustomI`. These are two possible combinations of key and value types. We can extend these types and scenario combinations as much as we want.

Step 3 – implementing a generic test run function

The next step required is to write a test helper function that contains all the logic for looping over the map of test cases and asserting the functionality of the `GetSortedValues` function. Usually, this functionality is contained inside the test body, but we will extract a function, as we have multiple input maps that we need to process:

```
func runTests[K ~int, V comparable](t *testing.T, tests
map[string]testCase[K, V]) {
    t.Helper()
    for name, rtc := range tests {
        tc := rtc
        t.Run(name, func(t *testing.T) {
            keys := make([]K, 0, len(tc.input))
            for k := range tc.input {
```

```
        keys = append(keys, k)
      }
      sort.Slice(keys, func(i, j int) bool {
        return keys[i] < keys[j]
      })
      sortedValues, err := gs.GetSortedValues(tc.input, gs.ASC)
      require.Nil(t, err)
      require.NotNil(t, sortedValues)
      for index, v := range sortedValues {
        key := keys[index]
        assert.Equal(t, tc.input[key], v)

      }
    })
  }
}
```

The runTests function has the following implementation:

1. The function signature takes two type parameters for the key and value types. They have the same type constraints as the test case type we have declared. It takes in two parameters: the *testing.T runner and the map of test cases, with the same key and value type parameters.

2. This function is marked as a helper using t.Helper so that it does not pollute test output.

3. We loop over the map of test cases and run each in its own subtest using the t.Run function.

4. The rest of the function body is the same as any test implementation. We invoke the function under test and assert the order of the elements.

Step 4 – putting everything together

With all our building blocks in place, we can write our first generic test for the GetSortedValues function:

```
func TestGetSortedValues(t *testing.T) {
  t.Run("[int]string", func(t *testing.T) {
    testStrings := map[string]testCase[int, string]{
      // values declaration
    }
    runTests(t, testStrings)
  })
```

```
    t.Run("[CustomI]float64", func(t *testing.T) {
      testFloats := map[string]testCase[CustomI, float64]{
        // values declaration
      }
      runTests(t, testFloats)
    })
  }
```

The test contains a subtest for each combination of key and value types. Inside the subtest, we create the map of test cases and invoke the runTests function with the correct parameters.

Step 5 – running the test

The last step is to run the test to see our generic code in action:

```
$ go test -run TestGetSortedValues ./chapter11/generics/sort -v
=== RUN    TestGetSortedValues
--- PASS: TestGetSortedValues
  --- PASS: TestGetSortedValues/[int]string
    --- PASS: TestGetSortedValues/[int]string/unordered
    --- PASS: TestGetSortedValues/[int]string/empty_map
    --- PASS: TestGetSortedValues/[int]string/negative_values
  --- PASS: TestGetSortedValues/[CustomI]float64
    --- PASS: TestGetSortedValues/[CustomI]float64/unordered
    --- PASS: TestGetSortedValues/[CustomI]float64/empty_map
    --- PASS: TestGetSortedValues/[CustomI]float64/negative_
keys PASS
ok  github.com/PacktPublishing/Test-Driven-Development-in-Go/
chapter11/generics/sort          0.201s
```

The usage of subtests gives us structured output for our test run. We have also easily adapted table-driven testing to use generics to be able to test our GetSortedValues function with different types of input values.

As we have seen from this simple example, the power of generics extends to test code. Another interesting use of generics is for creating test utilities. We will look at an example of this in the next section.

Test utilities

The test assertion libraries we have been using so far, `testify` and `ginkgo`, were written before the introduction of Go generics. They use reflection to implement comparisons and assertions on variables. While this is a powerful tool, it can be difficult to use to write our own assertions and test utilities. The introduction of generics has made this process much easier, so we can easily create our own test utilities to reuse them in our tests.

> **Reflection in a nutshell**
>
> Reflection is the ability of a program to analyze and manipulate program elements at runtime. This gives us the ability to inspect types, modify behavior, and test code. The Go standard library provides the `reflect` package, which provides functionality for implementing reflection.

Continuing our example from `TestGetSortedValues` we wrote previously, we can streamline our `runTests` helper by extracting a test utility that we can use in other tests. We can easily create our own generic test helper for asserting that the order of our keys and values is as expected:

```go
func AssertMapOrderedByKeys[K ~int, V comparable](t *testing.T,
input map[K]V, want []V) {
  t.Helper()
  keys := make([]K, 0, len(input))
  for k := range input {
    keys = append(keys, k)
  }
  sort.Slice(keys, func(i, j int) bool {
    return keys[i] < keys[j]
  })
  for index, v := range want {
    key := keys[index]
    assert.Equal(t, input[key], v)
  }
}
```

We extract the code and use two type parameters for the key and value type. We use the same technique as writing our `runTests` helper, but extract only the code that asserts on map value ordering. This test helper still takes the test runner `*testing.T` as a parameter, allowing it to fail tests in the case that the assertions inside it fail.

In the absence of generics, we would have to use the empty interface, `interface{}`, to allow our test to take in a variety of parameters. This does not allow us to write type-safe code, so writing helpers is more difficult and error-prone.

Generics can help us streamline our application and test code. As we have seen in this section, we can use it to extend the technique of table-driven testing to allow us to write tests that cover a variety of input types and scenarios. We have also seen how generics can be used to create our own test utilities.

Extending the BookSwap application with generics

So far, we have seen how to write a generic function and use generics to write easier test utilities. This has already proven to be a very powerful mechanism, providing us with both flexibility and type safety, something which cannot be achieved by an empty interface. In this section, we will learn how to make use of generics in our example REST API, the `BookSwap` application.

Let us suppose that the `BookSwap` application wants to extend its business model and begin swapping magazines, alongside its regular books business model. *Figure 11.3* presents the new system diagram for the application:

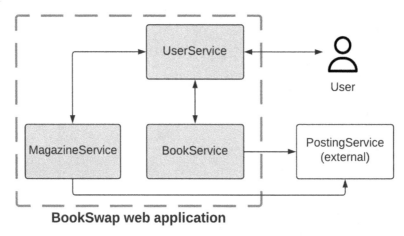

Figure 11.3 – The extended BookSwap application

The preceding example considers the `BookSwap` application's monolithic architecture, but the same kind of considerations would apply to microservices architectures as well. Changes would have to be made throughout the application to support a new model, starting from the database level:

- A `Magazines` database table will be created. Just like the `Books` table, it will have a foreign key dependency on the `Users` primary key, `id`.

- `MagazineService` will be created to interact with the database queries. Just like `BookService`, it will support upsert, list, and swap operations.

- `UserService` will have a dependency on `MagazineService`, allowing it to perform operations on this service and send information onward to the user.

- `PostingService` will need to handle both `Magazine` and `Book` when they are successfully swapped. Since this service is external, we can assume that this information will be transferred through an HTTP request.

Some of these changes do require dedicated code since we wouldn't want to make magazines and books too tightly coupled. One example of where we might leverage generics is during the construction of the HTTP response. So far, `Response` has consisted of only the `Books` items:

```go
type Response struct {
    Message string     `json:"message,omitempty"`
    Error   string     `json:"error,omitempty"`
    Books   []db.Book  `json:"books,omitempty"`
    User    *db.User   `json:"user,omitempty"`
}
```

We populate the `Books` slice with the books returned from `BookService`. We now need to extend the `Response` struct to be able to handle `Magazines` as well. `Response` is a widely used type, so it would be a good candidate for a generic implementation.

We create a `ResponseItemType` custom constraint, which contains a set of the `db.Book` and `db.Magazine` types:

```go
type ResponseItemType interface {
    db.Book | db.Magazine
}
```

If more types are added, then we can add them to this custom type constraint and use them throughout the application.

Next, we use `ResponseItemType` as the type parameter of `Response`:

```go
type Response[T ResponseItemType] struct {
    Message string     `json:"message,omitempty"`
    Error   string     `json:"error,omitempty"`
    Items   []T        `json:"items,omitempty"`
    User    *db.User   `json:"user,omitempty"`
}
```

We use the placeholder, T, as the type of the response, which is then used as the type of the Items slice. The Items slice is now able to contain both the db.Book and db.Magazine types. All other functions that interact with Response will now also need to handle the generic Response.

We add the same type parameter to the writeResponse function, which is responsible for populating data in Response:

```
func writeResponse[T ResponseItemType](w http.ResponseWriter,
status int, resp *Response[T]) {
    // implementation
}
```

The generic function simply passes the type to the response and no other changes are required for the implementation logic. The type either needs to be supplied or sent through as a placeholder.

On the calling side, we also need to handle the generic aspects. Each handler makes use of the writeResponse function to populate data on Response and return it to the client. The ListBooks handler demonstrates how to handle this on the calling side:

```
// ListBooks is invoked by HTTP GET /books.
func (h *Handler) ListBooks(w http.ResponseWriter, r *http.
Request){
    books, err := h.bs.List()
    if err != nil {
        writeResponse(w, http.StatusInternalServerError,
&Response[db.Book]{
            Error: err.Error(),
        })
        return
    }
    // Send an HTTP status & the list of books
    writeResponse(w, http.StatusOK, &Response[db.Book]{
        Items: books,
    })
}
```

The handlers responsible for books will handle response writing in a similar way. We pass the db.Book type to Response and invoke the writeResponse function. We do not need to pass a type argument to this function, as the type can be inferred from the invocation on the Response parameter. In the case of an error, we write the error to Response and return it, stopping execution. In the happy path case, we write the books to the Items slice.

The `Magazine` handlers will be implemented in the same way using the `db.Magazine` type instead. We can use the same table-testing technique we have explored in previous sections for testing our response logic.

This brings us to the end of our exploration with generic code in Go. This powerful tool allows us to write flexible code, which can be used with different data types. When it comes to generic code, we should always keep in mind that it needs to be tested against different types of input parameters, not just different values. This can make testing more complicated, but we can still easily modify the popular technique of table-driven testing to test generic code.

Testing best practices

This brings us to the end of this chapter and the end of our exploration of Go with TDD. We have had an amazing journey, having studied many libraries and techniques and applied them to our `BookSwap` application, as well as smaller examples. In this final section, we will recap the lessons learned and formulate some testing best practices.

Achieving a comprehensive testing strategy requires constant effort and a cultural shift within an organization to embrace quality and prioritize testing. *Figure 11.4* presents a summary of some of the best practices we have explored divided into three different categories:

Development	Testing	Culture
+ Use TDD to write tests during development process	+ Write simple, isolated tests	+ Document customer requirements
+ Use interfaces to wrap dependencies	+ Use table-driven tests to cover a variety of scenarios	+ Mitigate for unexpected errors and outages
+ Refactor code to tackle technical debt	+ Write tests at every level of the application	+ Value and prioritize code quality
+ Write robust code which is able to handle a wide variety of inputs	+ Use Go's `testing` package	+ Document non-functional requirements
+ Adopt generics for easier code reuse	+ Use third-party testing libraries	

Figure 11.4 – Development, testing, and culture best practices

We have discussed 14 best practices throughout this book, which have been divided into 3 categories – development, testing, and culture.

Development best practices

Here are some best practices for development (as seen in *Figure 11.4*):

1. **Use TDD to write tests during the development process**: The best way to ensure that the code we write is tested is to include it in the development process. Code is never delivered untested and developers write testable, well-designed code.

2. **Use interfaces to wrap dependencies**: Our code will often have dependencies on other components. It is a good practice to have dependencies external to our package wrapped by interfaces so that they can be easily replaced, either by test code or with another dependency. We have explored dependencies and interfaces in *Chapter 3, Mocking and Assertion Frameworks*.

3. **Refactor code to tackle technical debt**: Code should be refactored alongside new feature development to ensure that it stays performant, readable, and easy to maintain. We explored some refactoring techniques in *Chapter 7, Refactoring in Go*. Tests will ensure that no functionality is broken by the refactoring process.

4. **Write robust code that is able to handle a variety of inputs**: We discussed what code robustness is in *Chapter 10, Testing Edge Cases*. It should be able to handle a wide variety of inputs and respond with well-formulated errors.

5. **Adopt generics for easier code reuse**: As we have seen in this chapter, generics allow us to write code that is able to handle different data inputs. This allows us to use reusable code, which works for a variety of type parameters.

Testing best practices

Here are some best practices for testing (as seen in *Figure 11.4*):

1. **Write simple, isolated tests**: Tests should be simple and isolated from their dependencies. This allows us to avoid setting up a large number of services, which can be cumbersome and are likely to change. Instead, we should write focused tests that make use of mocks to test the UUT in isolation. We explored mocks and dependencies in *Chapter 3, Mocking and Assertion Frameworks*.

2. **Use table-driven tests to cover a variety of scenarios**: The popular technique of writing table-driven testing is to easily create a list of test cases and run through them. Test cases should be run in their own subtests to create a well-structured test output. We explored this technique in *Chapter 4, Building Efficient Test Suites*, as well as this chapter.

3. **Write tests at every level of the application**: Unit tests are fast, but they only assert that the functionality of a given package is correct, not that it is able to function correctly with other units. As depicted by the testing pyramid presented in *Chapter 1, Getting to Grips with Test-Driven Development*, we should write automated tests that assert that the individual units of the application integrate and function correctly together.

4. **Use Go's testing package**: While it might appear overly simple at first, the `testing` package offers a wide variety of functionality that allows us to write functional and non-functional tests. We have explored the capabilities of this package throughout this book.

5. **Use third-party testing libraries**: We explored multiple third-party testing libraries (`testify`, `ginkgo`, and `godog`) that complement the standard `testing` package and make it easier for us to write test assertions and create mocks.

Culture best practices

Here are some best practices for culture (as seen in *Figure 11.4*):

1. **Document customer requirements**: User journeys and customer requirements should be at the heart of all the tests we write. As it is nearly impossible to write tests that cover every single code path and interaction, engineers should ensure that the things that matter to customers are prioritized and covered by tests.

2. **Mitigate for errors and unexpected outages**: In microservices architectures, it is nearly impossible to ensure that there are zero outages. We should contract-test our services, as well as design our microservices architectures according to the best practices described in *Chapter 8, Testing Microservice Architectures*.

3. **Value and prioritize code quality**: As an organization, you should value and prioritize code quality, allowing the engineering teams time to refactor their services to ensure they can be easily extended and maintained for the future needs of the business.

4. **Document non-functional requirements**: While one important use of tests is to ensure that our system is able to satisfy functional requirements, another important aspect is to verify the performance of our system according to the needs of the customers. We explored how to test the performance of our code using benchmarks in *Chapter 2, Unit Testing Essentials*.

The final aspect to remember is that no testing strategy is perfect, as it cannot cover every single code path and edge case. The tools and techniques we have explored throughout this book should help you plan and implement the efficient testing of your system. There is no "one-size-fits-all" approach to a testing strategy, so make sure that you work with your product manager and other key business stakeholders to ensure that your testing efforts are made efficiently.

Summary

In this chapter, we spent time exploring Go's newly introduced generics support. We learned the basic syntax for implementing generic code, including how to specify type constraints. We also looked at a quick comparison of generic code versus writing code using interfaces.

Then, we revisited the previously introduced technique of table-driven testing and learned how to modify it to support generic code, allowing us to write test cases that support different input types and values. We also learned how to make use of generics to easily write our own test utilities, promoting code reuse in test code as well as implementation code.

Finally, we summarized all of the tools and techniques we have explored throughout this book with 14 best practices divided into 3 categories: development, testing, and culture. Implementing and maintaining a comprehensive testing strategy requires effort throughout the entire product and engineering organization.

Questions

1. What is generics?
2. What are the main components of Go generics?
3. Describe the steps involved in writing generic table-driven testing.

Further reading

- *Generic Data Structures and Algorithms in Go: An Applied Approach Using Concurrency, Genericity and Heuristics*, by Richard Wiener, published by Apress
- *Cloud Native Go: Building Reliable Services in Unreliable Environments*, by Matthew Titmus, published by O'Reilly

Assessments

Chapter 1, Getting to Grips with Test-Driven Development

1. The testing pyramid specifies how automated test suites should be structured. At the bottom of the pyramid are unit tests, which test a single isolated component. Next up in the middle of the pyramid are integration tests, which test that multiple components are able to work together as specified. Finally, at the top of the test pyramid are end-to-end tests that test the behavior of the entire application.

2. Functional tests cover the correctness of a system, while non-functional tests cover the usability and performance of a system. Both types of tests are required to ensure that the system satisfies the customers' needs.

3. The red, green, and refactor TDD approach refers to the three phases of the process. The red phase involves writing a new failing test for the functionality we intend to implement. The green phase involves writing enough implementation code to make all tests pass. Finally, the refactor phase involves optimizing both implementation and testing code to remove duplication and come up with better solutions.

4. Acceptance test-driven development. Just like TDD, ATDD puts tests first. ATDD is related to TDD, but it involves writing a suite of acceptance tests before the implementation begins. It involves multiple stakeholders to ensure that the acceptance test captures the customer's requirements.

5. Code coverage is the percentage of your lines of code that are exercised by your unit test. This is calculated by considering the function statements, parameter values, and execution paths of your code. The Go test runner outputs the calculated code coverage. We should aim for a good value, but optimizing for 100% is normally not appropriate.

Chapter 2, Unit Testing Essentials

1. In Go, a module is a collection of packages that should be built and released together. A package is a collection of Go files that must be built together. A module is specified by a `go.mod` file, while a package is specified by the package declaration at the top of the source file.

2. The additional test package is a package that matches the name of the source package with the `_test` suffix added. The additional test package provides isolation between test code and source code, preventing brittle tests and allowing developers to see the integrations with their packages.

3. Test signatures have only a few requirements:

    A.   The test name must begin with the word `Test`

    B.   Any other name that follows the `Test` must also be capitalized

    C.   There must only be one parameter of the `*testing.T` type and no return type

4.   Subtests are tests that run within an enclosing test. They are created by passing a name and corresponding function to the `t.Run` method on the enclosing test. A failure in a subtest will cause the failure of the enclosing test as well.

5.   Benchmarks are a way to measure the performance of our code. Go's test runner will repeatedly run the instrumented function until it finds stable measurements for the performance of the benchmarked function.

Chapter 3, Mocking and Assertion Frameworks

The code examples included in this book are publicly available at `https://github.com/PacktPublishing/Test-Driven-Development-in-Go`. You can find the implemented solution from the Question section in the `chapter03-solution` directory.

Chapter 4, Building Efficient Test Suites

1.   An edge case is a test case that occurs at extreme values of an operating parameter. A corner case occurs at the extreme values of multiple operating parameters. As corner cases occur when multiple edge cases occur, they are much less likely to occur than edge cases.

2.   An idempotent operation is an operation that can be repeated multiple times without changing the initial result. These operations are predominant in API design, which can often involve retries and resending requests due to network conditions.

3.   In Go, error handling is done using the built-in `error` type. Errors are handled as part of regular code flow, where errors are handled just like any other return value, most often alongside other values using multiple return values.

4.   Table-driven testing is a popular technique that allows us to test multiple scenarios in a unified way, which reduces code duplication. Custom types are created, which represent the inputs and expected outputs of the particular test scenario. These are then saved in a test collection. The test collection is run over, with each case being identically executed.

5.   By default, Go tests in different packages are run in parallel. Test cases within a package can be marked to be run in parallel using the `t.Parallel()` method of the `*testing.T` instance. This allows tests that are marked accordingly to be run in parallel.

Chapter 5, Performing Integration Testing

1. Integration tests cover multiple units of the system, ensuring that they work well together. End-to-end tests replicate user behavior on the system as a whole.

2. BDD is a branch of TDD that focuses on writing human-readable tests and involve a variety of stakeholders from across the business. The most important part of BDD is establishing a common vocabulary that is shared across the business.

3. No, it is considered an anti-pattern to mock databases. They are complex systems that are difficult to mock and replicate. In general, tests should use the same database as in production.

4. A container is a unit of software and all its dependencies, specified by a container image. Docker Engine starts up and manages containers on top of the operating system.

Chapter 6, End-to-End Testing the BookSwap Web Application

1. A user journey is a path that a user may take to achieve their goal while using a given application. Identifying user journeys allows us to write E2E tests that correctly replicate and verify the user experience on our platform.

2. ORM stands for object-relational mapping and is a technique that allows us to bridge the gap between object-oriented languages and relational databases. It allows developers to interact with the database types as just any other custom type.

3. Docker Compose allows us to easily define and network multi-container applications. Dockerfiles define the steps to building a single container image, while Docker Compose allows us to define services with a single configuration and command.

4. Database seeding involves adding initial data to a database. Often, this dummy data is generated and changes between test runs.

Chapter 7, Refactoring in Go

1. Code redesign involves changing the functionality of an existing service, while code refactoring involves changing the internal structure of an existing service, without changing its existing functionality. Code redesign changes the *what*, while code refactoring changes the *how*. If done correctly, code refactoring will not be visible outside the refactored code scope.

2. The developer starts by identifying the change they wish to make. Then, they modify their implementation or test code. Once this is done and the code compiles, they then run their test suite. If the suite is passing, the code refactor has been successful. However, if the suite is not passing, the developer revisits their code change, adjusting it together with the test code until the test suite is passing. No change is released until the test suite successfully verifies the change.

3. Technical debt is the term used to refer to code that has not been written to the technical team's typical high standards of quality. It typically occurs when the team prioritizes speedy delivery over taking the time to design a high-quality solution.

4. A monolithic application is a single application that is built and released as a single unit. A microservice architecture is a system design pattern that involves building independently built and released units. Organizations typically start with monolithic applications and then move to microservice architectures as needed.

Chapter 8, Testing Microservice Architectures

1. Functional testing ensures that the features of the system work correctly. Non-functional testing verifies that other aspects of the system behave as expected. The two main types of non-functional testing are performance tests and usability tests.

2. Performance testing relies on key metrics to quantify and compare the performance of the application. Important key metrics to monitor are response time, error rate, concurrent users, data throughput, and CPU/memory usage.

3. Performance testing ensures that the system is scalable by measuring the performance of the individual parts of the system, allowing us to identify bottlenecks and improvements required. Furthermore, testing at higher loads of the system allows us to estimate the limits of what the system can handle, and helps us understand what the growth runway of the current system configuration is.

4. Microservice architectures provide scalability benefits because they allow the different parts of our system to be scaled independently. They also allow easier maintenance and provide increased delivery speed, as engineering teams can own and change their services without any central oversight. On the other hand, they introduce complexity in every part of the software life cycle

    I. Development complexity as engineers need to structure multiple code bases, then implement and test features across multiple services.

    II. Deployment complexity as each service has its own infrastructure to be maintained. Automation of the release pipelines becomes essential in order to make it easy for engineers to release multiple services.

    III. Organizational complexity as engineers must handle increased ownership of multiple services. While teams are able to deliver features on multiple services without any central oversight, they must also communicate to establish common technical standards, suitable for a variety of microservices.

5. Contract testing is a testing technique that allows us to reliably test the interaction between two services. The consumer initiates the exchange and sends the request, while the provider processes the request and sends back the response. The exchange is recorded in a contract on the consumer side. Then, the contract is verified on the provider side against the real implementation. One of the key advantages of contract testing is that it allows us to verify the integration between two services without having to run and maintain the test against real implementations of both the consumer and provider.

Chapter 9, Challenges of Testing Concurrent Code

1. Concurrency refers to a program's ability to execute more than one task, with interruptions and without any ordering guarantees. Parallelism refers to a program's ability to execute more than one task, simultaneously and without interruptions. The OS (or even in silico implementations such as hyperthreading) may give the illusion of parallelism through pre-emptive multitasking. However, in order for parallelism to be truly simultaneous, multiple computational resources are required.

2. Channels support three operations. The send operation writes information to the channel, while the receive operation reads information from the channel. The close operation signals to all receivers that no more values will be sent through it. Once closed, channels can never be reopened. For unbuffered channels, send and receive operations are synchronous and will only complete once both sender and receiver are available. On closed channels, sends will panic, while receives will immediately complete with the zero value of the channel's data type.

3. `sync.Mutex` is a lock that exposes the `Lock` and `Unlock` methods. Goroutines will block until they can acquire the lock successfully. `sync.WaitGroup` is a specialized lock that maintains an inner counter that will block until the inner counter of `WaitGroup` reaches 0 by calling `Done` on `WaitGroup` the appropriate number of times. `sync.WaitGroup` allows us to wait for multiple goroutines to complete, in a simpler and more compact way than reading from channels or shared memory.

4. A data race occurs when multiple goroutines access and modify a shared resource. This can lead to inconsistent values and hard-to-detect bugs. A deadlock occurs when goroutines are blocked, waiting for a resource that never becomes available. If the whole program becomes blocked, the Go runtime will detect the deadlock and shut down the program.

5. Go's race detector is integrated into the go tool chain and is enabled with the `-race` command-line flag. It instruments memory access and reports when data races are detected. As it significantly increases the CPU and memory usage of the applications it instruments, the race detector should not be used in production.

Chapter 10, Testing Edge Cases

1. A robust system is one that continues to function correctly, even when supplied with unexpected or exceptional inputs. When errors or unexpected inputs occur, the system is able to handle them gracefully without panics and return meaningful errors. Robust code is readable, maintainable, and easy to test.

2. Fuzz testing is a software testing technique that involves generating a large amount of random data, which is then passed to the fuzz target in an attempt to verify its behavior against a wide range of parameters. Fuzz testing makes it easier to uncover bugs by reducing the number of tests we need to write manually to cover edge cases of input variables.

3. A fuzzed test begins with the `Fuzz` prefix, takes in a single `*testing.F` parameter, and returns no values. Just like other tests, they must be defined inside `_test.go` files. They are run using the `go test` command.

4. Property-based testing is a technique that involves the generation and verification of input values that satisfy a set of system properties or specifications. This allows us to test a wide variety of edge cases that are relevant to our system without the need to manually write and maintain a large number of tests.

5. From the viewpoint of the UUT, untrusted inputs are any values that the UUT receives or that it uses that it has not generated. These inputs can be good candidates for fuzzing arguments.

Chapter 11, Working with Generics

1. Generics is the ability to write code that is able to be used with different data types without being limited to a single type. It is a powerful mechanism, supported by most programming languages, and allows us to write flexible, reusable code.

2. The main components of Go generics are type parameters, type constraints, type arguments, and type inferences:

    I.   Type parameters are the placeholder types that will be used in our generic code, typically denoted by a single letter.

    II. Type constraints restrict the types that can be used with our generic function or type.

    III. Type arguments are the types that are passed to the function upon invocation. They replace the type parameters and satisfy the type constraints.

    IV. Type inference is the process that takes place under the hood to determine the type of a variable, without us needing to explicitly declare it. This makes generic code less verbose.

3. Testing generic code with table-driven tests will often involve writing test cases with different type parameters. To support this, we define generic test case types that can contain different input types. Then, a generic test run function that is able to run over the generic test cases and interact with the input types should be created. Test cases should be run in subtests, ensuring that test runs have a structured output.

Index

Symbols

packtpub.com

Subscribe to our online digital library for full access to over 7,000 books and videos, as well as industry leading tools to help you plan your personal development and advance your career. For more information, please visit our website.

Why subscribe?

- Spend less time learning and more time coding with practical eBooks and Videos from over 4,000 industry professionals

- Improve your learning with Skill Plans built especially for you

- Get a free eBook or video every month

- Fully searchable for easy access to vital information

- Copy and paste, print, and bookmark content

Did you know that Packt offers eBook versions of every book published, with PDF and ePub files available? You can upgrade to the eBook version at packtpub.com and as a print book customer, you are entitled to a discount on the eBook copy. Get in touch with us at customercare@packtpub.com for more details.

At www.packtpub.com, you can also read a collection of free technical articles, sign up for a range of free newsletters, and receive exclusive discounts and offers on Packt books and eBooks.

Other Books You May Enjoy

If you enjoyed this book, you may be interested in these other books by Packt:

Domain-Driven Design with Golang

Matthew Boyle

ISBN: 978-1-80461-345-0

- Get to grips with domains and the evolution of Domain-driven design
- Work with stakeholders to manage complex business needs
- Gain a clear understanding of bounded context, services, and value objects
- Get up and running with aggregates, factories, repositories, and services
- Find out how to apply DDD to monolithic applications and microservices
- Discover how to implement DDD patterns on distributed systems
- Understand how Test-driven development and Behavior-driven development can work with DDD

Event-Driven Architecture in Golang

Michael Stack

ISBN: 978-1-80323-801-2

- Understand different event-driven patterns and best practices
- Plan and design your software architecture with ease
- Track changes and updates effectively using event sourcing
- Test and deploy your sample software application with ease
- Monitor and improve the performance of your software architecture

Packt is searching for authors like you

If you're interested in becoming an author for Packt, please visit `authors.packtpub.com` and apply today. We have worked with thousands of developers and tech professionals, just like you, to help them share their insight with the global tech community. You can make a general application, apply for a specific hot topic that we are recruiting an author for, or submit your own idea.

Share Your Thoughts

Now you've finished *Test-Driven Development in Go*, we'd love to hear your thoughts! Scan the QR code below to go straight to the Amazon review page for this book and share your feedback or leave a review on the site that you purchased it from.

https://packt.link/r/1-803-24787-8

Your review is important to us and the tech community and will help us make sure we're delivering excellent quality content.

Download a free PDF copy of this book

Thanks for purchasing this book!

Do you like to read on the go but are unable to carry your print books everywhere?

Is your eBook purchase not compatible with the device of your choice?

Don't worry, now with every Packt book you get a DRM-free PDF version of that book at no cost.

Read anywhere, any place, on any device. Search, copy, and paste code from your favorite technical books directly into your application.

The perks don't stop there, you can get exclusive access to discounts, newsletters, and great free content in your inbox daily

Follow these simple steps to get the benefits:

1. Scan the QR code or visit the link below

https://packt.link/free-ebook/9781803247878

2. Submit your proof of purchase
3. That's it! We'll send your free PDF and other benefits to your email directly

www.ingramcontent.com/pod-product-compliance
Lightning Source LLC
Chambersburg PA
CBHW062057050326
40690CB00016B/3129